MW01352755

Journey to Civilization

The Science of How We Got Here

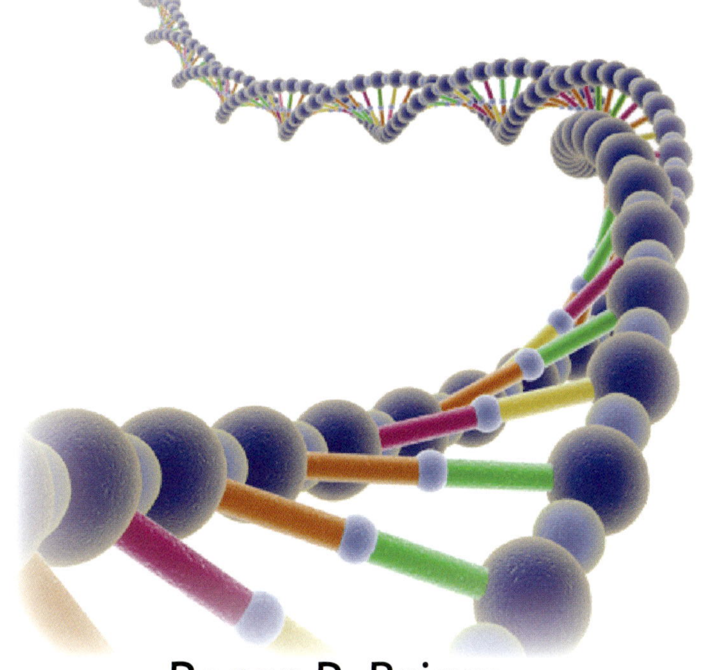

Roger P. Briggs

Collins Foundation Press

Journey to Civilization

Published by the Collins Foundation Press
4995 Santa Margarita Lake Road
Santa Margarita, CA 93453
www.CollinsFoundationPress.org

Managing and Production Editor - Cheryl Genet
Copy Editor - Vera Wallen

Cover design by David Jordan Williams

Includes biographical references

ISBN 978-0-9884382-0-0

Printed by Sheridan Books in the United States of America

Dedicated to

My science mentor,
Albert A. Bartlett,
Professor of Physics at the University of Colorado
and

my father,
William E. Briggs
1925 - 1999

CONTENTS

DETAILED CONTENTS

Note: In the text, mya = million years ago, ya = years ago

Preface

My main motivation for writing *Journey to Civilization* came from two places: a life-long love of science and a deep curiosity about our origins, that is, how we humans got here. Regarding the first of these, I must acknowledge that many people today do not think of themselves as lovers of science, perhaps because they were confused or intimidated in some uninspiring science class. Yet science is one of the most natural of human endeavors, stemming from our innate curiosity about the world. We can experience a sense of awe and even joy when we learn about the secrets of nature. It is my hope that this book will help readers rediscover that sense of wonder and curiosity about the world that we had as children.

The second of these, the need to know about our origins, is something that is deeply rooted in human culture and consciousness. By about 100,000 years ago, ancient people were beginning to acquire spoken language and telling stories, passing them on from old to young, down through the generations. These mythical traditions, the stories of "our people," were important in all human societies because they provided a common ground of meaning by explaining to people how the world was made, how people came to be, and what human life means. Myths gave people a sense of order and place in the universe.

However, in the last few hundr=ed years, we have lost our creation stories as modern science has compelled us to question their validity and diminished their cultural power. Now our stories, our myths, are about sports teams and charismatic media stars, or cowboys clearing out the Indians so America could be built, or the lives of people on television. But these stories cannot satisfy our need to know our origin and place in the universe. We have been left disconnected and adrift, as we casually exploit and ravage the Earth.

Since the late twentieth century, a new kind of origin story has been emerging from the discoveries of science. *Journey to Civilization* tells this new story, the story of the universe and life and humanity, based entirely on mainstream, well-accepted science. But unlike every origin story before, this new story is universal: it is the origin story of all the people of the Earth. It establishes a common ground for all of us, regardless of nationality, religion, race, or any other difference among us; and it points toward a new sense of our place in the universe.

There were two pervasive challenges that I faced in writing this book, aside from the monumental task of covering nearly 14 billion years of history. First was the question of how much depth and detail to go into with each new part of the story and each new area of science. Too much depth would make this book thousands of pages long and lose all but the most hardcore science geeks. Too little would trivialize the magnificent knowledge that we have amassed and the achievements of the remarkable scientists who devoted their lives to discovery. So I tried to strike a balance, and I apologize right up front to the many scientists who will feel that there was so much more that could be said about their field of expertise.

The second challenge, more a frustration, comes from the fact that scientific knowledge is constantly changing and expanding. The pace of scientific discovery and the rate of growth of our knowledge have been continuously accelerating for the last century, and one has to wonder how long this can continue, even though there are no signs of this slowing down. For this reason, a book of this kind will be out of date in some ways from the moment it is published, no matter when. We plan to publish revised and updated versions but it is my hope that, in the bigger scheme of things, the origin story according to science will be a project that humanity will continue to take up and improve upon. This book is a contribution to that project.

I would like to acknowledge some of the people who generously offered their time and energy to improve this book. I am grateful to Ron Biela, Beth Bennett, Bill Briggs, MaryAnn Briggs, Scott Brown, Chip Chace, Jeff Etter, Beverly Hackenberg, Chip Lee, Dan McBride, and Scott Winston for reading my manuscript at critical stages, and for their constructive suggestions.

Thanks also to astronomer Roger Linfield for helping me get the astronomy and astrophysics right; to paleoanthropologist Robert Corruccini for his suggestions on Hominid evolution; to Eric Miller for inspiration and support throughout my science teaching career, and for access to his incredible fossil collection; to Mark Bekoff for his encouragement and guidance at times when I really needed it; to Russ Genet for his thorough reading of my manuscript and his many suggestions to strengthen it; to Joanne Ernst for editing, counsel, and support throughout the project; to Vera Wallen for her keen eye and sense of clarity in late stage editing; to Cheryl Genet for believing in this project from just about the moment she saw it; to Collins Foundation Press and Cheryl Genet for turning my work into a book; and to Jim Collins and Jon Krakauer for their ongoing inspiration and encouragement.

Roger Briggs
Boulder, Colorado

Journey to Civilization

INTRODUCTION

What is "The Science of How We Got Here"?

Whether stuck in traffic or surfing the internet, most of us are so immersed in present-day civilization that we never stop to reflect on how incredible it is that human civilization exists on this blue-green paradise we call Earth, a planet that teems with life and circles a benevolent star in a quiet neighborhood of the Milky Way galaxy. How *did* we get here?

People were asking this question long before anyone knew what planets and stars and galaxies were. We have always needed to know how the world was made and how people came to be. For tens of thousands of years we relied on creation stories to answer these existential questions, to make sense of a frightening world, and to explain the meaning and purpose of human life. In the ancient Babylonian story, the world was created by Marduk who was himself the son of gods, and Marduk created humans to labor in the world and to worship the gods—this was the purpose of human life. For the Hopi people of the American southwest, there was in the beginning only the Creator, Taiowa, who made Sotuknang and entrusted him with making the nine solid worlds that included Earth and its people. And the Old Testament Bible says that God created the Earth and its people in six days and that he made people in his own image.

Cultural creation stories such as these have deeply satisfied people for eons with explanations of the origins and meaning of life. The many different stories and beliefs that can be found worldwide demonstrate the great variety of cultures in our world. While some creation stories share common themes from culture to culture, there has never been one story shared by all the people of the world. Today, however, nearly all of humanity shares the methods and products of science. We rely heavily on science for our explanations and answers. After all, it has brought tremendous power and prosperity to people everywhere, ranging from nuclear weapons and mobile phones, to artificial limbs, cures for deadly diseases, and the internet. We have arrived at a juncture in time where we have the exciting opportunity to explore the ancient question of how we got here from the *scientific perspective*—a perspective that people everywhere can choose to share, as science has become a universal language in today's world.

As recently as the mid-twentieth century, science could do very little to explain how we got here. It could not explain what life is, or how it got started on Earth; it could not account for how and when the Earth and Sun were made; it had no theory of how the universe itself was born and how stars and elements and planets are created; and it could not say how long humans have been on Earth and what they were doing for most of that time. For most of the last 300 years, science could only tell us that we were tiny, insignificant specks in a virtually infinite universe. And modern people have been increasingly disconnected from nature, with little sense of place or purpose in the universe.

However, in the later twentieth century, spectacular discoveries in many fields of science, particularly molecular biology, astrophysics, and paleoanthropology, made it possible for the first time to seriously consider our origins. The dazzling pace of scientific discovery has continued unabated into the twenty-first century, and now a new kind of creation story is emerging for humankind—one that springs entirely from the evidence and skepticism of science.

In 2005, after a 30-year career as a classroom science educator, I began the project of finding the creation story according to science. I wondered how well science could account for the existence of human civilization on Earth, and I wanted to learn and know the science behind the story. So I set out in search of the science of how we got here, of how the universe was born and how it produced a Sun and Earth, how life began and slowly turned into upright-walking creatures with very large brains, how these *Homo sapiens* developed technology and language and eventually spread over the entire Earth, and how human civilization finally emerged. This book is the result of my search.

Journey to Civilization is not just the story of how we got here, but also an exploration of the science behind the story, and how we know that science. To understand this story we must journey through many fields of modern science including astrophysics and cosmology, evolutionary biology, molecular genetics, and paleoanthropology. Explanations are written for the curious *non-scientist*, yet some readers may tend to feel overwhelmed or intimidated by new terms and technical-sounding jargon. But please be assured, however, that you do not have to master every concept and term before you can move ahead with the story. Simply soak it in, revel in the depth and richness of human knowledge, and marvel at what we have achieved. In the end I hope you will change your view of humanity.

Dealing with Time

The story of how we got here is set in time, and time presents many challenges as we try to understand it. Even without considering the more baffling aspects of time that Einstein discovered, such as the warping of time in strong gravity, we will find that simply comprehending the time scales in the history of life and the history of the universe is very difficult.

To illustrate this, consider these four time spans that are of major importance in our story:

5000 years – the length of time civilization has existed.

200,000 years – the length of time humans (*Homo sapiens*) have existed.

4.55 billion years – the length of time the Earth has existed.

13.75 billion years – the length of time the universe has existed.

It is very difficult to grasp these time spans, even the shortest one. "History," in the traditional sense, goes back only about 5000 years to the first civilizations of Sumeria when writing was invented, and until recently everything before that has been called "prehistoric." It is just about all we can do to imagine how long ago it was that the Romans lived, and the Greeks before that, and the Egyptians before that, but this takes us back only about 4,000 years. To understand our deepest history—the origins of humanity, and the history of life, the Earth, and the universe—requires a completely different view of time that is often called *geologic time*. Geologic time, with its *hundreds of millions* of years and *billions* of years, is so vast and deep that no one can really comprehend it. Yet we will need to deal with it.

Another problem with time is the *relative size* of the different time spans we will be talking about. For example, how does the 5000-year history of civilization compare to the 13.75 billion-year history of the universe? One way to convey the answer is to say that if we used the distance of *one mile* to represent the age of the universe, then the age of civilization would be only about *half a millimeter*, or the thickness of a few sheets of paper. Put another way, it's like the thickness of a human hair on a 100-meter football field. You cannot look at both things at the same time. Yet this is what we will be trying to do.

To help with these problems, we will use a *geocosmic timeline* like the one shown on page 5. This timeline divides the 13.75 billion-year life of the universe into three equal intervals—we will call these the three *geocosmic days*. A geocosmic day is simply the length of the time that the Earth has existed, and we know (as of 2003) that the universe is just about exactly three times as old as the Earth. The idea of the geocosmic day is not so different from the other natural time units we use:

One day = the amount of time it takes for the Earth to rotate once on its axis.

One year = the time it takes for the Earth to circle the Sun.

One geocosmic day = the amount of time the Earth has existed; that is, the age of the Earth.

Using the geocosmic day as a unit of time will be helpful in several ways. First, we get a better understanding of how old the universe is when we think of its life as filling <u>three</u> *geocosmic days*. This means that the universe is almost exactly three times as old as the Earth. Most of the life of the universe (*two-thirds* of it) occurred *before* the Sun and Earth even existed! Secondly, because most of the action in our creation story takes place on the Earth, filling only the third geocosmic day, clock times during that third day will help give us a better time perspective on important events. For example we can say that the beginning of life on Earth, roughly 3.8 billion years ago, happened at about 4:15 in the morning of the third geocosmic day; that dinosaurs went extinct at about 11:40 pm (at night!) on that last day, and humans finally showed up at about 11:59:56 pm—just 4 seconds before midnight. A more complete listing of important events in the creation story is given next page. As we move through billions of years in the next twelve chapters, we will often represent the life of the universe in three geocosmic days. This helps us get a handle on time and the vast time scales of our story.

This book is structured in five parts, representing five great ages in the story of the universe. Part One covers the first two-thirds of the life of the universe, the time before the Sun and Earth existed, from the birth of the universe, through the emergence of stars and galaxies, to the formation of the Sun and Earth. Parts Two and Three tell the story of life on Earth, from its mysterious origins to the age of large mammals and primates. Parts Four and Five explore the labyrinthine saga of the hominid lineage, from the development of upright walking, to the appearance of *Australopithecus* and *Homo*, the emergence of humanity, and finally civilization.

If we look at the geocosmic time scale representing the life of the universe and the list of events next page, it is clear that we humans are a very, very recent arrival. The entire hominid lineage, going back about 7 million years, fills only the last two minutes of the third geocosmic day in the life of the universe. Modern humans do not show up until about 4 seconds before midnight (which represents now), and civilization finally kicks-in at 11:59:59.9 pm—one-tenth of a second before midnight! That is to say, the entire field of traditional western history—going back 5000 years to the first city states in Sumeria, then Egypt, Greece, Rome, and Europe—would fill only the last *one-tenth of a second* in the three-day life of the universe. Until recently we knew almost nothing about the rest of that time, before the first civilizations. But now, our discoveries and growing knowledge in many fields of science make it possible to tell the whole story, the big story, of how we got here, from the very beginning. We better get started.

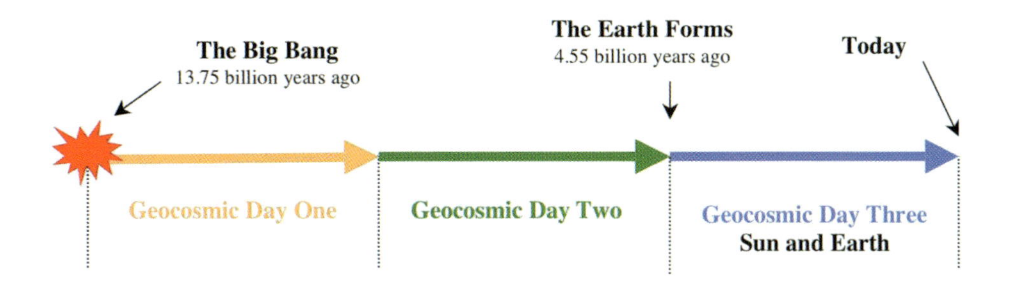

A Geocosmic Timeline of the Universe
Here the life of the universe is depicted in three geocosmic days.
One geocosmic day equals 4.55 billion years, which is the age of Earth.

Main Events in the Story of the Universe
(All dates are approximate)

Event	Actual Years Ago	Geocosmic Time*
The Big Bang	13.75 billion	Day One, 12:00 am (midnight)
The first stars	13.3 billion	Day One, 2:00 am
The Milky Way Galaxy forms	12.7 billion	Day One, 5:15 am
Dark energy begins to dominate	7.5 billion	Day Two, 8:00 am
Formation of the Earth	4.55 billion	Day Three, 12:00 am (midnight)
Life begins	3.8 billion	Day Three, 4:15 am
First complex cells (eukaryotes)	2.4 billion	Day Three, 11:15 am
The Cambrian explosion of life	540 million	Day Three, 9:00 pm
Age of the dinosaurs	240-65 million	Day Three, 10:40-11:40 pm
First human-like ancestors (genus *Homo*)	2.5 million	Day Three, 11:59 pm
First anatomically modern humans	200,000	Day Three, 11:59:56 pm
Civilization begins	5,200	Day Three, 11:59:59.9 pm
Today	0	Midnight

**One second of geocosmic time = 52,800 years of real Earth time*

Part One

The Age of the Cosmos

PART ONE

THE AGE OF THE COSMOS
Building a Perfect Planet

The world we know came into existence long, long ago at a moment that we have come to call the *Big Bang*. Scientists sought for much of the twentieth century to find out exactly when the Big Bang happened, but not until the year 2003 was there overwhelming evidence that this moment occurred 13.75 billion years ago.* This was the birth of the observable universe and the beginning of the story of how we got here. Nine billion years after this explosive birth, our star and planet emerged from the cosmos. In Part I of our story we will see how the cosmos grew from a single point that was infinitely hot and dense into the elegant and beautiful universe we now see and the blue-green planet that nurtures and protects complex life.

---

* See Appendix I for an explanation of how we found the age of the universe.

Time Context for Part One

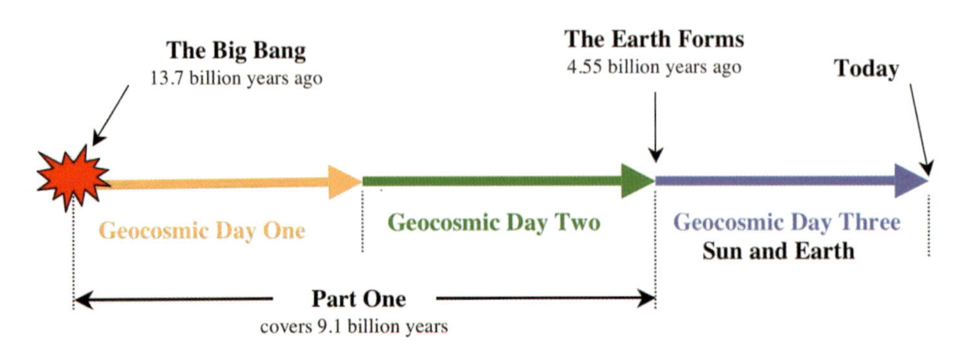

Main Events of Part One: The Age of the Cosmos

Event	Time after Big Bang	Geocosmic Time
First nucleons	1.0 second	Day one, 12:00 am
First neutral atoms	380,000 years	Day one, 12:00:07 am
First stars	400 million years	Day one, 2:00 am
First galaxies (including our own, the Milky Way)	400 - 600 million years	Day one, 2:00-3:00 am
Dark energy begins to dominate over gravity	6 billion years	Day two, 8:00 am
The Sun forms	9.0 billion years	Day two, 11:30 pm
The Earth forms	9.1 billion years (4.55 billion years ago)	Midnight between day two and day three.

Chapter 1
The First 380,000 Years

In the Beginning

In the beginning, when the universe was born, all was one. Matter, energy, space, time, and anything else we can name, did not exist separately but were all merged together in a pure oneness, the tiniest seed that would grow into the complex and elegant universe we now live in. We have no physics or mathematics that can describe this moment, this originary state of cosmic unity when everything was compressed into a single point that was infinitely hot and dense; scientists call this state a *singularity*.

From this singularity, as time itself came into existence, the universe was born in a gigantic detonation that hurled everything outward, triggering a relentless expansion that we can still see today. The starting temperature and density of the universe were nearly infinite, but as expansion set in and the universe grew in size, the temperature and density dropped steadily; the universe has expanded and cooled for its entire life.

As the infant universe expanded and cooled, it slowly took on the structure of the world we know today: a world made of quarks and electrons, of atoms and molecules, that gathered into stars and galaxies and planets; a world with at least one planet where molecules assembled into something called life; a world where life, in one location at least, continued for so long that it evolved into intelligent, reflective, self-aware beings who now ponder all this.

Exploring Deeper

An Inevitable Question

Can we ask what was happening *before* the Big Bang, before the beginning of the universe? Some scientists feel that something may indeed have been happening before our universe was born—a previous universe, perhaps, that ended in a crushing *Big Crunch*. Or, maybe our universe is but one of many universes that exist independently of ours—the *multiverse* scenario that some mathematical theories predict. Still, it seems very unlikely that we will ever know anything about what happened before our universe began because the nearly infinite temperatures and densities of the Big Bang would surely have destroyed any remnants of information from before. It seems certain that no information could survive the Big Bang—it was the ultimate sterilizing event that eliminated all traces of anything that preceded it. Our universe must have begun with an absolutely clean slate, a completely unknown past. So, while we *can* ask what was happening before the Big Bang, it seems that we may never be able to know the answer.

Science and Discovery

The Two Great Pillars

How is it possible to find the moment when the universe began, to date the Big Bang? How does science go about finding the answer to that question? Who are these scientists on such a quest, and what are their tools of discovery? They are astronomers following in the footsteps of Galileo, who first used the telescope in 1609 to gaze upward and probe the hidden secrets of the heavens. In more recent years they have become astrophysicists, combining the observations and data from astronomy with the tools of physics to understand what's out there—from stars and galaxies, to black holes and planets. And among physicists, astronomers, and astrophysicists are specialists called *cosmologists* who study the universe as a whole, seeking to know how and when it was born, how it became the world we know, and what its future may hold. For more details about how scientists built on Galileo's early discoveries to find and date the birth of the universe, see Appendix I.

Today the most widely accepted theory of the universe among cosmologists is called the *Hot Big Bang Theory*. It is built on the mathematical predictions of the two great pillars of physics, *quantum mechanics* and *general relativity*, aligned with the data that astronomers gather from their ever-more powerful telescopes. The field of quantum mechanics deals with the very smallest things in the world—atoms and nuclei and subatomic particles. In contrast, the field of general relativity describes gravity and the very large-scale universe. Each of these theories was developed in the early twentieth century, but quite differently: quantum mechanics was first developed over more than ten years, culminating in 1927, by a stellar collection of physicists and mathematicians that included Werner Heisenberg, Erwin Schrödinger, Niels Bohr, Paul Dirac, and Max Born, while the general theory of relativity was created almost entirely by just one man, Albert Einstein, and first published in 1915.

However, these two great theories of the world, relativity and quantum mechanics, have never been compatible with each other—they make conflicting predictions in some cases. One of the great dreams of modern physicists is to somehow merge the two into one unified theory that has been dubbed *quantum gravity*. A more complete understanding of the early universe awaits this merger, yet even without a theory of quantum gravity, cosmologists have been able to construct a likely chronology of the early universe that is supported by recent observations. We know that by the time the universe was one second old, many very important events had already taken place.

Much Ado About One Second

The first *one second* was the busiest time in the entire existence of the universe. A number of crucial events must have taken place by the time the universe was one second old, and cosmologists have created an elaborate and detailed chronology of events that occurred during this time (see Figure 1-1 for a summary).

Figure 1-1
The First Moments in the Life of the Universe

Name	Time after the Big Bang	What was happening
The Planck Epoch	10^{-43} seconds	Unknown
Grand Unification Epoch	10^{-35} seconds	*Gravity*, then the *strong nuclear force* separate. Quarks form.
Inflationary Epoch	10^{-34} to 10^{-32} sec	Universe expands in size by at least 10^{25} and quantum fluctuations amplify. Electrons and neutrinos form by the end of this time.
Electroweak Epoch	10^{-12} seconds	The *electromagnetic* and *weak nuclear forces* separate, making four fundamental forces. Neutrinos separate (decouple) from matter and radiation can stream out into space.
Hadron Epoch	10^{-6} to 1.0 seconds	Free quarks join to form protons and neutrons (hadrons). This is the birth of "normal" matter. Matter and anti-matter meet and annihilate, leaving only matter.

For the first fraction of a second after the moment of the Big Bang—for just 0.0 000000000000000000000000000000000000000001 (or 10^{-43}) seconds—nothing is known except that the size of the currently visible universe was virtually zero while the density and temperature were nearly infinite. Cosmologists call this the *Planck Epoch* and say that it cannot be understood until a theory of quantum gravity emerges. However, in the next phase, just after the Planck Epoch, the first recognizable features of our universe began to appear. This has been called the *Grand Unification Epoch* and is marked by the emergence of gravity, the strong nuclear force, and the fundamental particles of matter known as *quarks*. And when this work was done, about 10^{-34} seconds into the lifespan of the universe, something preposterously unimaginable apparently took place; that something is now referred to as *cosmic inflation*.

The *Theory of Cosmic Inflation* was put forward around 1980 by two cosmologists working independently, Alexei Starobinsky at the Landau Institute in Moscow and Alan Guth at MIT in Boston. They were trying to explain how the universe could be smooth and homogeneous in its infancy and later take on a large-scale structure that was anything but smooth, made up of galaxies and clusters of galaxies, with enormous voids in between. Inflationary theory asserts that at a very short time after the Big Bang (10^{-34} seconds), the universe grew almost instantly by a factor of 10^{25} or more, driven by a mysterious repulsive force that is thought to permeate everything, even empty space. This mysterious force that hyper-inflated the infant universe has variously been called *vacuum energy* and *zero-point energy,* and it may be related to *dark energy,* which we will encounter in Chapter Two.

Quantum theory predicts mathematically that from the beginning, tiny *quantum fluctuations* existed, and then cosmic inflation caused these miniscule variations to grow astronomically to become the framework for the large-scale universe. Astronomers have recently seen this structure for the first time in the "bubbly" web-like patterns that galaxies and galaxy clusters trace out on scales of billions of light-years (see Figure 2-3 on pg. 30).

When cosmic inflation ended at about 10^{-32} seconds, the stage was set for the remaining fundamental forces and particles to "freeze out" as the temperature of the universe kept dropping. Before this, things were so hot that all matter and energy, and the fundamental forces, were merged into one. After gravity and the strong nuclear force appeared, the remaining two forces emerged by about 10^{-12} seconds after the Big Bang. These were the *electromagnetic force* and the *weak nuclear force.*

During the same time, the remaining fundamental particles of matter also formed. Particle physicists have now identified 12 fundamental particles, divided into two groups, called *quarks* and *leptons* (see Figure 1-2). The six quarks are named *up, down, top, bottom, strange,* and *charm* (arbitrary meaningless names), and each has a companion anti-quark. Among other things, quarks combine to make protons and neutrons, which themselves are the building blocks of normal, familiar matter. The six leptons include the familiar *electron* and the elusive *neutrino.* As with the quarks, each lepton has an anti-particle. The commonplace electron, for example, has an anti-matter version called the *positron* that is rarely seen today.

When a particle of *matter* meets its *anti-matter* counterpart, they completely annihilate each other, turning all the mass into a large amount of energy. Although the annihilation of matter and anti-matter almost never happens in today's universe—except in the exotic experiments of particle physicists—it was common when the universe was approaching the age of one second. In a cosmic shootout, particles of matter and anti-matter met and annihilated, producing a prodigious flash of radiation that still lingers today. By the time the universe was one second old, the shootout was over and matter had prevailed over anti-matter because there was slightly more of it to begin with. The slight bit of surviving matter became the visible universe we see today.

Now that the universe was about one second old, things were cool enough for quarks to combine in threes to form protons and neutrons. This was the birth of normal matter, with protons and neutrons becoming the building blocks of the world we know. The hydrogen nucleus—a single proton—now existed, but it was still far too hot for electrons to attach to that nucleus and form neutral hydrogen atoms. Heavier elements like carbon and oxygen formed much later, when protons and neutrons could fuse together into larger nuclei, but this would not begin for another 400 million years, when the first stars were born.

Figure 1-2
Building Blocks of the Universe: The Standard Model of Particle Physics

Particles That Make Up Matter

The Six LEPTONS	(and their electric charge)	
light	**electron** (-1)	**electron neutrino** (0)
↓	**muon** (-1)	**muon neutrino** (0)
heavy	**tau** (-1)	**tau neutrino** (0)

The Six QUARKS	(and their electric charge)	
light	**up** (+2/3)	**down** (-1/3)
↓	**charm** (+2/3)	**strange** (-1/3)
heavy	**top** (+2/3)	**bottom** (-1/3)

Particles That Transmit the Fundamental Forces

FUNDAMENTAL FORCE	PARTICLE THAT CARRIES IT
Electromagnetic	**Photon**
Strong Nuclear	**Gluon**
Weak Nuclear	**W, Z Vector Bosons**
Gravity	**Graviton**

Science and Discovery

The Particle Zoo
The strange-sounding zoo of particles that make up everything was proposed and discovered not by cosmologists, but by another group of physicists working the opposite end of things: particle physicists. While cosmologists consider the largest scales of nature—the universe as a whole—particle physicists explore the smallest scales of nature, searching for the tiniest particles that make up everything that is.

However, cosmology and particle physics merge together in the Big Bang theory because this is when the universe *and* the fundamental particles both came into existence.

Unlike cosmologists, particle physicists can actually do experiments to probe the composition of matter. They use particle accelerators to do this. These are machines that can accelerate particles like protons up to very high speeds, close to the speed of light, and then smash them together. In the aftermath of the violent collision, a shower of new particles emerges. This is how scientists discovered many of the fundamental particles that make up matter. As bigger and bigger accelerators have been built, new particles have been discovered. The collision of very high energy-particles is essentially a simulation of the early universe.

The largest particle accelerator today is called the Large Hadron Collider or LHC, located near Geneva, Switzerland. The LHC is an underground circular tube, 27 kilometers around, which accelerates two beams of protons in opposite directions up to about the speed of light; it then directs them together to make a collision with such energy and density that it approximates the first moments after the Big Bang. The LHC was switched on for the first time in the fall of 2008, but because of its enormous complexity it was plagued with startup problems until the spring of 2010 when it was successfully run at half power. When fully operational, it promises to answer some of the lingering questions about the deepest structure of matter and the birth of the universe.

In the 1970s, particle physicists proposed an organizing scheme that has come to be called the *Standard Model of Particle Physics*, summarized in Figure 1-2. Many particles, like the quarks, were predicted theoretically before they were actually detected. Since 1964 when the quark was first proposed and named by Murray Gell-Mann (and scoffed at by many of his peers), particle accelerators have reached higher and higher energies and, one by one, all six quarks have been detected with the attributes that were predicted from theory. We now know that the *proton* is made of three quarks, two *ups* and one *down*, giving it a total charge of +1, while the *neutron* is made of one *up* quark and two *downs*, giving it a charge of zero.

One of the longstanding mysteries of particle physics, and a missing piece of the Standard Model, has been the elusive Higgs boson that was first proposed in 1964 by Peter Higgs and others. They theorized that this particle endowed the other fundamental particles with mass. For forty-eight years researchers debated its existence and searched for signs of it, but to no avail. One of the great hopes for the Large Hadron Collider was that it would reveal the Higgs, and indeed in July of 2012 physicists working with the LHC cautiously announced that they had found it. Although little is yet known about the Higgs boson, its confirmation puts the Standard Model on much firmer ground.

The Rules of the Game

Something else must have emerged very early in the life of the universe: the rules of the game, so to speak. Physicists call these rules the *Laws of Nature*. They describe how nature works. The main reason that science is so powerful is that we have discovered laws of nature and can use them to predict and control events. One of the primary laws of nature was proposed by Isaac Newton in the late 1600s and is known as the *Universal Law of Gravitation*. It can be summed up in a simple equation that tells how gravity works:

$$F = \frac{G m_1 m_2}{R^2}$$

This says that any two pieces of mass ($\mathbf{m}_1$ and $\mathbf{m}_2$) that are a distance **R** apart will *attract* each other with a force (**F**) that we call the gravitational force. This is the force that holds you down to Earth—the same force that you measure when you stand on a bathroom scale. The size of that force—really, the strength of gravity—is determined by a special number symbolized as **G** in the equation. Physicists refer to this special number as "big G". **G** is a *fundamental constant of nature* and its value is known very accurately (see Figure 1-3). We do not know why it has exactly this value, but we do know that if the value of **G** was anything else, the large-scale universe would be very different. If **G** was a little bigger or a little smaller, then gravity would be stronger or weaker, and the large-scale universe, from galaxies right down to the Earth, would have formed very differently. It seems that the precise value of **G** has been crucial in making the universe work just how it does.

Figure 1-3
Some of the Most Important Fundamental Constants of Nature

Name	Symbol and Value	What it Relates to
Gravitational Constant	$\mathbf{G} = 6.67 \times 10^{-11}$ N-m^2/kg^2	Gravity
Permittivity of Free Space	$\varepsilon_0 = 8.85 \times 10^{-12}$ C^2/N-m^2	Electricity
Permeability of Free Space	$\mu_0 = 4\Pi \times 10^{-7}$ Wb/A-m	Magnetism
Speed of Light in Vacuum	$\mathbf{c} = 3.00 \times 10^8$ m/s	Light
Boltzmann's Constant	$\mathbf{k}_B = 1.38 \times 10^{-23}$ J/K	Temperature and energy
Planck's Constant	$\mathbf{h} = 6.6 \times 10^{-34}$ J-s	Atoms and photons

In the three centuries since Newton discovered how gravity works, we have uncovered the workings of the other great phenomena of nature: electricity, magnetism, light, temperature and energy, the atom, and the nucleus. There are laws (usually equations) that describe how each of these work, and each phenomenon has a fundamental constant associated with it, just as **G** is associated with gravity (Figure 1-3). How the universe turned out, and the way it works, is determined by the precise values of these fundamental constants. Some scientists have called this *fine tuning,* meaning that the fundamental constants seem to be fine tuned to make the universe just exactly what it is. The values of the fundamental constants, and the laws of which they are part, must have been set very early in the life of the universe. By the time the universe was about one second old, matter had formed, the rules of the game were set, and the great drama of existence could begin to unfold.

Exploring Deeper

Big Bang or Big Bounce?

The cover of the October 2008 *Scientific American* proclaims, "Forget the Big Bang: Now it's the Big Bounce." The author of the cover story, Martin Bojowald, heads a team of cosmologists at Pennsylvania State University that has pioneered a new theory called *loop quantum gravity* (Bojowald 2008). One of the features of this theory is the existence of tiny particles of spacetime, akin to the particles of matter we call atoms.

These spacetime atoms would be completely unnoticeable under most circumstances, but when spacetime is packed closely together, as it was in the very early universe, they become very significant. The theory predicts that at extremely high densities these particles of spacetime cause gravity to become a *repulsive* force rather than an attractive force. This would mean that a true singularity could not exist because this repulsive gravity would prevent things from getting that small and dense. This repulsive gravity caused the early universe to expand until densities got low enough that gravity switched over to being the attractive force we are familiar with. The repulsive gravity that acted when densities were extremely high could have been the driver of cosmic inflation.

The existence of repulsive gravity when the universe was very small and young suggests the possibility of a previous universe that was collapsing under attractive gravity—a *Big Crunch*—until it got so small that repulsive gravity kicked in to stop the collapse and turn it around into expansion. This was the *Big Bounce*. If this is true, it would mean that our universe is just one of perhaps an infinite string of universes, collapsing, bouncing, and expanding endlessly, eternally.

In their early work with the equations of loop quantum gravity, Bojowald and his team hoped they would be able to extrapolate backward from the birth of our universe to a previous universe, thus dispensing the notion of "the beginning of time." Further work suggested that during the bounce, when density was nearly infinite, the universe passed through an "unfathomable quantum state" in which all information about a previous universe was scrambled and lost. It seems that if there was a previous universe that bounced into ours, we can never know about it. We are apparently stuck with knowing only about this universe we live in, which was born by Big Bang, or Big Bounce, 13.75 billion years ago.

Ancient Light: Microwave Soup

After the universe was one second old, things slowed down significantly. The next 200 seconds is called the period of *nucleosynthesis*, in which nearly all of the deuterium (a heavy form of hydrogen), helium, and a small amount of lithium formed. These were the first elements in the universe. Eventually, 100 or so elements would form and become the building blocks of the Earth and life.

During the 1940s George Gamow and others showed from theoretical physics that these first light elements could form only at extremely high temperatures—*billions of degrees*. Such temperatures could only arise in a scenario like the early Big Bang. Gamow predicted theoretically that the very hot conditions just after the Big Bang would produce a Universe that was made up of about 75% hydrogen, 24% helium, and traces of lithium. Every observation of the visible matter in the universe has so far confirmed these abundances, supporting the theory that the universe had to be almost infinitely hot at some point in its history.

After the period of nucleosynthesis, the universe continued to expand and cool into an opaque *plasma* of light nuclei, free electrons, and trapped photons (particles of light). The original source of these trapped photons of light was the intense flash that resulted from the annihilation of matter and antimatter when the universe was about one second old. However, this light was not free to travel through space—it was held captive in a dense mixture of matter and radiation as photons were absorbed and re-scattered continually by the free electrons. Finally, when the universe was about 380,000 years old and the temperature had dropped to about 3000 degrees, neutral matter could form; that is, electrons could attach to nuclei to form atoms. This was the birth of normal atomic matter. This transition from ionized matter (a plasma) to neutral matter (atoms) is called *recombination*;* it was one of the most significant moments in the infancy of the universe because photons of light could now escape from matter and travel freely through space. An enormous flash of ultraviolet and visible light filled the 380,000 year-old universe, and this light has persisted in space for the 13.7 billion years since. But in the time since this first flash of light was produced, the universe has been expanding, and because space itself is subject to this expansion, the wavelength of that original visible flash has been stretched into much longer microwaves (from about 10^{-7} meters to about 10^{-2} meters, a factor of about 100,000). Some of these photons that were liberated when the first neutral atoms formed have traveled for billions of years until being collected in the mirror of a telescope here on Earth. We see this light today as the *cosmic microwave background* (or the CMB), and the stretching from visible to microwaves is a clear signature of the expansion of the universe caused by the Big Bang.

Scientists today study the cosmic microwave background with great interest, using ever-more sensitive instruments to detect it, because they are seeing the light from the very early universe. It is truly the oldest light in the universe, giving us a snapshot of the universe in its infancy. The cosmic microwave background has become a rich source of information about the evolution and structure of the universe. It contains clues about the first moments after the Big Bang when it originated;

* *Recombination* is really a misnomer in this context since electrons and nuclei were *combining* to form atoms for the first time. The term comes from plasma physics and refers to the situation in which neutral atoms (with electrons and nuclei combined) are ionized to become a plasma, then recombine to become neutral atoms once again.

it records a snapshot of the period of recombination starting about 380,000 years ago when it was first released, it holds fingerprints of the first stars that ignited about 400 million years after the Big Bang, and it encodes evidence of its multi-billion year journey to Earth. Starting in about 1990 with the observations made by space-borne instruments like COBE, and later by WMAP (see the following Science and Discovery section, below), there is now abundant data supporting the theoretical models of the early universe; and not surprisingly, scientists have plans to deploy yet more sensitive instruments, such as the Planck Spacecraft, that will produce even better data.

The cosmic microwave background, and the vast amount of information it contains, has now become the smoking gun implicating the Big Bang as the birth of the universe. The evidence supporting the Big Bang now falls into three main areas:

- The redshift, or stretching, in the light from distant galaxies indicates that they are moving away from us. This means the universe is expanding, and suggests that it exploded at sometime in the past.* See Appendix II for more details on Doppler shift.
- The observed abundances of the light elements deuterium, helium, and lithium are consistent with the predictions of the hot Big Bang theory.
- The cosmic microwave background radiation and the corresponding background temperature of 2.73°K are remnants of the Big Bang and record that the Big Bang occurred 13.75 billion years ago.

* The redshift of galaxies is actually the weakest of the supporting evidence—the Steady State Theory (see below) can also explain the redshift, but it cannot explain the other two pieces of evidence.

Science and Discovery

The Cosmic Microwave Background

How can we make sense of the claim that the universe is flying apart? We see no evidence of it in our everyday lives. By 1915, Einstein had found mathematically from his newly developed *General Theory of Relativity* that the universe could either expand *or* contract. He knew that gravity should cause the universe to contract, but he and others at the time believed in a static universe that neither expanded nor contracted—this was ten years before Edwin Hubble discovered the expansion of the universe (see Appendix I). To counteract the shrinking effects of gravity, Einstein added a repulsive factor in his equations, thereby describing a universe that was static and unchanging. This factor later became known as the *cosmological constant.* but when Hubble discovered the expansion of the universe about ten years later, Einstein deleted this factor from his equations, and later called it his greatest blunder. It appears today that this was no blunder at all—the cosmological constant was resurrected in the 1990s when dark energy was discovered, and Einstein was probably right all along (more on this later in the chapter).

Others, like Alexander Friedmann and Georges Lemaitre, working in the 1920s, used the equations of General Relativity to support the idea of an expanding universe and suggested that this expansion implied a violent moment of birth for the universe long ago—a "big bang." When Hubble and others discovered the redshifts of galaxies (see Appendix I), the expansion of the universe was undeniable. This was, and still is, a shocking revelation; after all, the part of the universe in which we are located does not seem to be flying apart. But local gravity masks this expansion within our own galaxy and group of galaxies, and only when we look out at great distances can we see it.

The idea that the universe was born in a gigantic explosion was not immediately accepted. In 1948 Fred Hoyle, Hermann Bondi, and Thomas Gold proposed an alternative explanation for the expansion of the universe. Their *Steady State Theory* took the view that the universe was uniform in space and time; that is, it looked the same from every location, and had no beginning or ending. A universe with no beginning had a certain philosophical appeal because it disposed of questions about how it was created and what was happening *before* the beginning. To remain in this static or steady state, new matter would constantly be created to compensate for the out-flowing expansion. Their calculations showed that new matter would need to appear at a modest rate of about one atom per year in a 100-meter cube of space. Hoyle and the other adherents could not account for *how* this new matter was created or *where* it came from—it would just appear in empty space—but the effect would be so subtle that it *could* be happening without notice.

Meanwhile, in the 1940s and 50s, the *Big Bang Theory* was being refined by theorists like George Gamow and Ralph Alpher. Their work in the new field of nuclear physics using quantum mechanics showed that the very early universe, just after the Big Bang, had to be *very* hot in order to produce the first helium, and that an intense burst of light would have been produced shortly after the Big Bang. They predicted that this burst of radiation should still linger today as microwaves, stretched out by the expansion of the universe. Gamow and the other early Big Bang theorists made a very specific prediction that should have been measurable: they predicted that a faint soup of microwaves would fill all of space today, and that this *cosmic microwave background* would give the entire universe an overall background temperature of about 5 degrees above absolute zero. If this microwave remnant could be detected, it would be strong evidence that the Big Bang had actually occurred.

By the early 1960s, the search was on for the cosmic microwave background radiation. Robert Dicke and James Peebles, working at Princeton, began building small sensitive detecting antennas that could listen for this faint remnant of the Big Bang. But while they were having no success, just a few miles away at Bell Telephone Laboratories, Arno Penzias and Robert Wilson were experimenting with a new type of radio telescope that was sensitive to microwaves. They had become frustrated by a faint background hiss that remained no matter where they pointed their antenna. They simply could not get rid of it. But when they learned about the predictions of Gamow and heard about the work of Dicke and Peebles, they finally realized that the frustrating hiss they were picking up must be the cosmic microwave background radiation. Indeed, they had unintentionally detected it.

Their measurements, and others that soon followed, showed that the universe has an ambient temperature of 2.7 degrees, not far from Gamow's prediction of 5 degrees. Penzias and Wilson had indeed detected the redshifted aftermath of the Big Bang—the cosmic microwave background—and won the Nobel Prize in 1978 for this work. After their discovery was published, the Big Bang Theory became widely accepted and the Steady State Theory fell by the wayside.

Since the early 1960s when the microwave background was first detected, we have studied it in ever-greater detail with increasingly sensitive instruments. In 1989 the Cosmic Background Explorer (COBE) satellite was launched for this purpose. COBE carried three instruments that could make precise measurements of the cosmic microwave background (CMB) and when the first data streamed down, it showed a background temperature of 2.725 degrees above absolute zero (corresponding to microwave radiation) just as predicted by the Hot Big Bang theory. After several years of analysis by thousands of scientists from around the world, the data from COBE showed that the microwave background had a clear *anisotropy*—it was not completely smooth but had slight fluctuations when measured in different directions. This anisotropy in the CMB showed that the lumpy structure of the universe was well developed by the moment of recombination 380,000 years after the Big Bang. The 2006 Nobel Prize in Physics was awarded to John Mather from NASA's Goddard Spaceflight Center and John Smoot from the University of California at Berkeley for their roles leading to these discoveries. But these findings begged for more data from even more sensitive instruments.

In 2001 NASA launched the Wilkinson Microwave Anisotropy Probe (or WMAP), a space-borne instrument that could detect the microwave background with even greater precision—it measured tiny fluctuations in the background temperature to a *hundred-thousandth* of a degree. It was the early data from WMAP in 2003 that allowed scientists to pinpoint the age of the universe at 13.73 billion years (in 2010, after six more years of analyzing the data, the age of the universe was revised slightly to 13.75 billion years). The data that has arrived from WMAP since 2003 has shown more details of the microwave anisotropy and the foamy structure of the universe that had developed by the age of 380,000 years (Figure 1-4). This has provided support for the theory that cosmic inflation must have occurred just after the Big Bang (Cowen 2008).

Figure 1-4

A microwave picture of the universe taken by WMAP clearly showing the anisotropy, or foamy structure, of the cosmic microwave background.

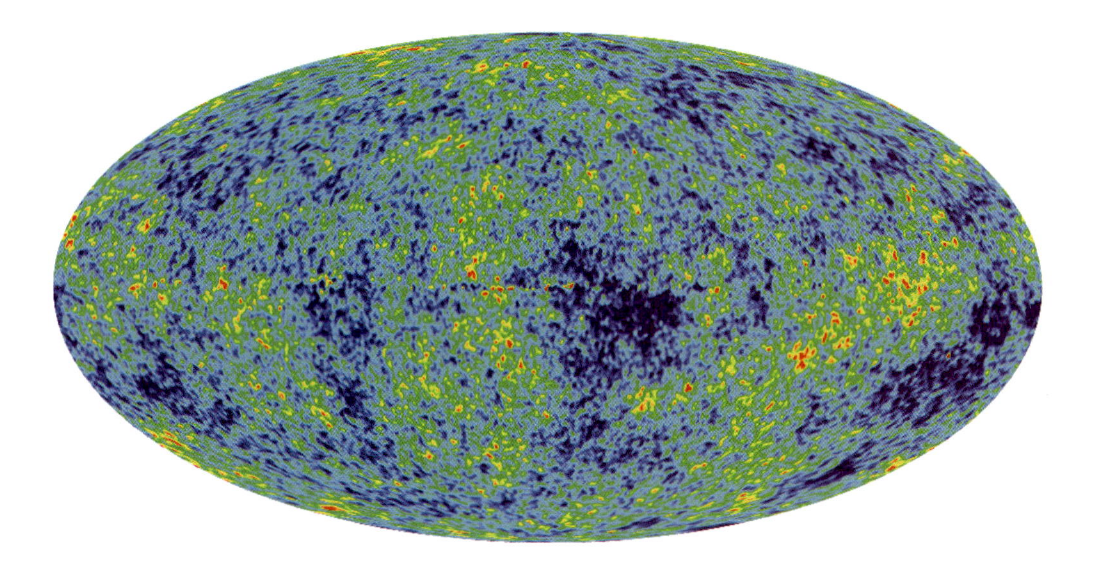

NASA/WMAP Science Team

Chapter 2
Large Scale Structure

Dark Matters: The Cosmic Web

Cosmologists have now discovered that something else must have formed in the earliest stages of the universe, something that has been hidden from us until very recently: *dark matter*. By about 1960, astronomers had noticed that the stars in some galaxies revolved around the center of the galaxy much faster than expected, as though gravity was much stronger than all the known matter could produce. This seemed to suggest that galaxies had much more mass than all of the stars and dust we could see (because they emit light). Calculations suggested that we were seeing only about 10% of the mass of galaxies and the other 90% was somehow invisible, or dark. Apparently, this dark matter was producing gravity, as all matter does, but not light.

More recent observations have hinted that the entire universe also has a great deal of "missing mass" or dark matter. The latest data suggest that the universe we can see—all of the stars and galaxies and clouds of gas and dust—may comprise only a few percent of the total contents of the universe. Apparently, most of the universe is invisible!

We do not yet know what dark matter actually is, only that it must be there and that it cannot be any form of *baryonic matter* (the familiar kind of matter made of protons and neutrons). This rules out things like burned out stars (called black dwarfs), cold dust clouds, and probably black holes. Some physicists now suspect that dark matter may be mostly comprised of some kind of exotic particle (dubbed the *WIMP*, for weakly interacting massive particle) that interacts only very weakly with normal matter except through gravity; but, as yet, this particle has not been detected (the Large Hadron Collider may eventually solve the mystery of dark matter).

The presence of dark matter is now being mapped out by its effect on things we *can* see—things that produce light. One method uses an effect called *gravitational lensing*, which is the bending of light as it passes through regions of strong gravity. In the last decade, gravitational lensing images of large sectors of the universe have revealed a gigantic skeleton of dark matter on which the visible universe hangs.

Another method of mapping out dark matter uses the distribution of galaxies at very large scales. These maps show that galaxies are not smoothly spread out, but concentrated in enormous sheets and filaments, with huge voids in between. This uneven distribution of galaxies traces out an intricate scaffolding of dark matter that has come to be called the *cosmic web* (see Figure 2-1) (Simcoe 2004).

Figure 2-1

Computer simulations of the cosmic web of dark matter, 2 billion years after the Big Bang (above) and today (below). The earlier box is 30 million light-years across while the later box is four times bigger, because of 12 billion years of cosmic expansion.

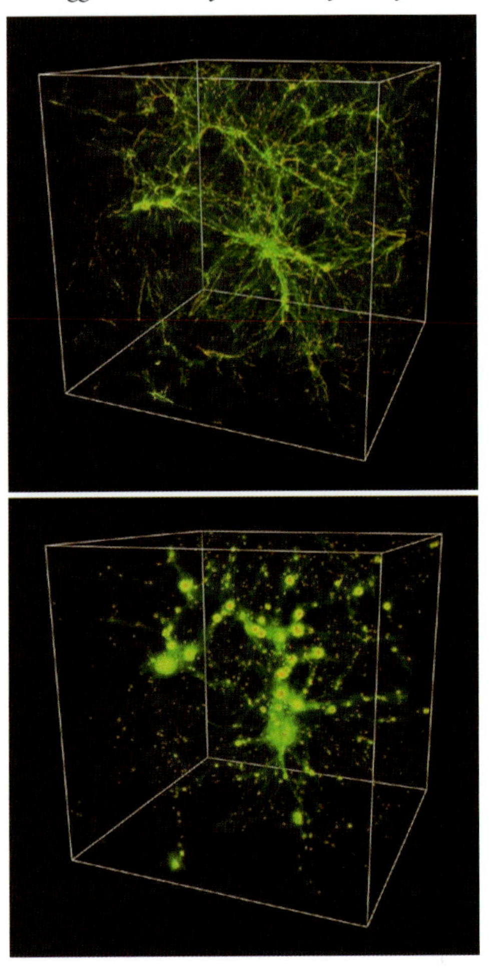

Courtesy: Renyue Cen, Princeton University

A third technique that reveals the cosmic web looks at the huge empty spaces between galaxies, which are not really empty but filled with a very thin gas of hydrogen and helium. When astronomers analyze the light that has passed through this *intergalactic medium,* they find that the gas is not spread out uniformly, but concentrated along the filaments of a web with varying density. All of the different methods used for detecting the cosmic web are now coming into agreement, allowing researchers to create maps of the cosmic web of dark matter and the large-scale universe. Although the cosmic web is completely invisible, its existence is now well established. But how did it form?

It is believed that dark matter arose during the first second after the Big Bang, along with the protons and electrons of normal matter. For the next 380,000 years it was too hot for the protons and electrons of the normal matter to combine into neutral atoms and release the photons they entrapped. So these protons and electrons and photons remained as a hot smooth plasma that produced an outward gas pressure. This outward pressure counteracted the effects of gravity, so it remained smooth and without structure until it cooled sufficiently. But the much more abundant dark matter followed a different path because it was not subject to an outward gas pressure. The effects of gravity, combined with the small density variations that grew out of the period of cosmic inflation, allowed the dark matter to begin clumping into bubbly filaments and sheets very early in the life of the Universe. This was the birth of the cosmic web, and it grew as the universe expanded, and thickened as gravity pulled more dark matter toward denser filaments.

When the Universe was 380,000 years old, the normal matter had cooled enough to become neutral atoms and the effects of gravity became dominant. By this time the cosmic web of dark matter was well formed and the normal matter quickly began to fall into the web. The large-scale structure of the Universe was now established, but it was nothing more than a veil of hydrogen and helium gas clinging to the cosmic web. The universe was still dark because stars did not yet exist; it was still too hot for them to begin forming.

At the age of about 400 million years, the universe had finally cooled enough for matter to begin falling into small dense clumps of dark matter to become the first stars (Bromm 2009). When the first stars ignited, the radiation that poured out from them ionized surrounding clouds of hydrogen and helium, causing them to glow. The faint glow of this *re-ionization* is the fingerprint of the first stars and has been detected indirectly in the details of the cosmic background radiation. The birth of the first stars was one of the most important events in the life of the Universe because stars perform some of its most crucial functions. Not only do they produce vast quantities of energy, they also manufacture elements and create planetary systems. We will look more closely at the birth, life, and death of stars later in this chapter. After many stars had formed, they began to fall into large regions of dark matter, sometimes called halos. This was the birth of the first galaxies.

Galaxies and Dark Energy

Galaxies are perhaps the most magnificent structures in all of nature, and also some of the oldest and most distant objects in the universe. The further away from us they are, the older they are, because the light from them that we see had to travel billions of years to reach us. That is, the further away a galaxy is, the further back in time we are seeing when we look at it. Observations of the most distant galaxies suggest that the first galaxies formed about 600 million years after the Big Bang, as the result of stars, dust, and gases being pulled into blobs of dark matter. As the matter in an infant galaxy contracts, it also begins to rotate. The more it contracts, the faster it rotates, like a spinning ice skater pulling her arms inward. The effect of this rotation is clear in the shape of large spiral galaxies (Figure 2-2).

Figure 2-2(a)

A galaxy seen "edge-on": The Sombrero Galaxy (M104) is about 30 million light-years away, and is moving away from us at about 640 miles per second! A supermassive black hole resides at the center of this galaxy.

NASA/STSci

Figure 2-2(b)

A galaxy seen "face-on": The Pinwheel Galaxy (M101) is about 27 million light-years away and about 170,000 light-years in diameter. Both the Pinwheel and the Sombrero Galaxy were first seen in 1781.

NASA/STSci

Figure 2-2(c)

Andromeda (M31), our sister galaxy, is very much like our own galaxy, the Milky Way. The smaller bright spots above and below the center are among the many dwarf satellite galaxies in our Local Group. Andromeda is about 2.5 million light-years away and about 100,000 light-years in diameter.

Courtesy: Robert Gendler: http://www.robgendlerastropics.com/

The first galaxies were small and disorganized, but some were drawn toward other infant galaxies by gravity, merging to form larger galaxies. The first few billion years in the life of the universe were dominated by mergers of smaller galaxies into larger ones, and some eventually grew into massive spiral galaxies like our own Milky Way. It is believed that our galaxy is one of the oldest and was probably beginning to form about 13 billion years ago, or about 700 million years after the Big Bang. Galaxies also formed in groups and clusters of groups, guided by the filaments and nodes of the cosmic web of dark matter that permeate the large-scale universe. Around every galaxy is a much larger halo of dark matter that is merely a tiny bulge in a filament of the web.

The Milky Way and its companion Andromeda are large spiral galaxies and are surrounded by at least twenty smaller galaxies, forming what has come to be called the *Local Group*. Recent observations indicate that Andromeda and the Milky Way are moving toward each other and will eventually merge into one super-galaxy. Our Local Group is part of a much larger cluster of galaxy groups, the *Virgo Supercluster*. A map of our neighborhood of the universe that is one billion light years across is shown in Figure 2-3. The presence of the cosmic web is very apparent at this large scale.

Figure 2-3

A map showing galaxy clusters in our neighborhood of the Universe, about one billion light-years across. Our cluster, our galaxy, our star, and our planet all sit near the center in the Virgo Supercluster.

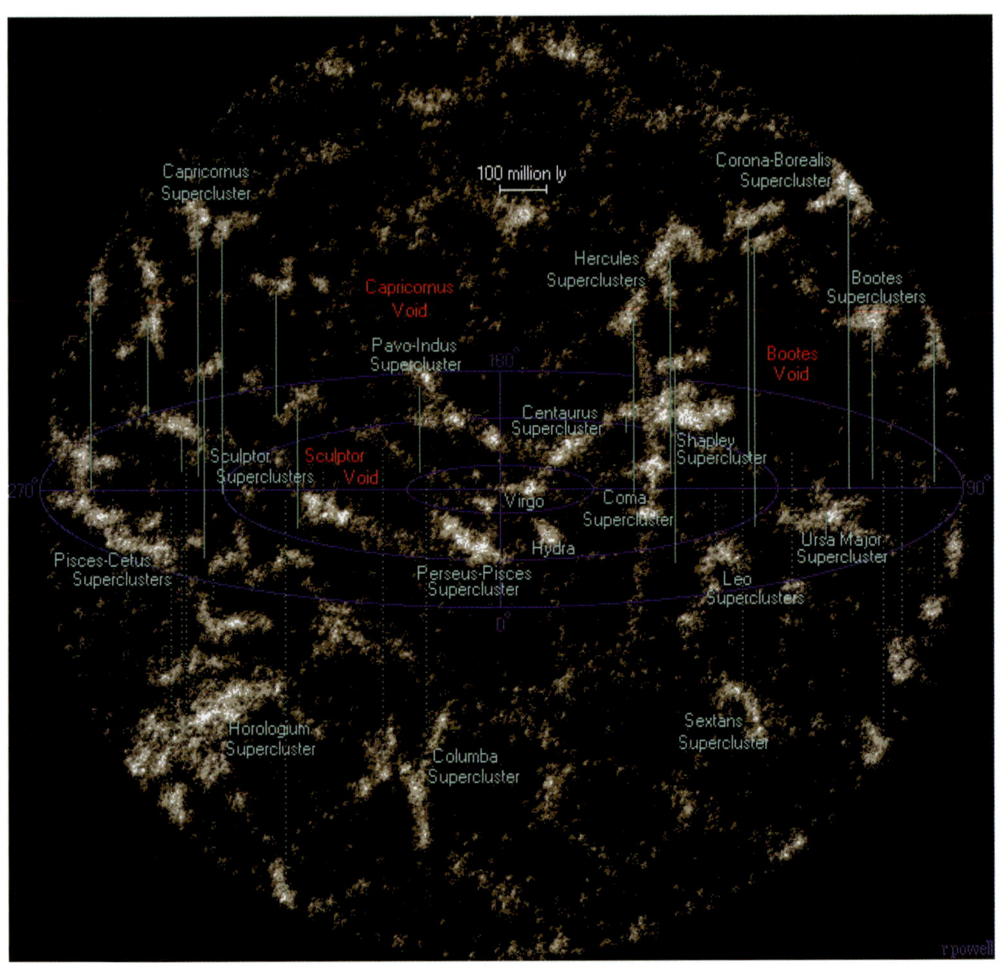

Courtesy: Richard Powell, http://www.atlasoftheuniverse.com

Astronomers have been able to study the formation of early galaxies by looking out at great distances with instruments like the Hubble Space Telescope. The most distant galaxy seen to date, and thus one of the oldest, is 12.9 billion light-years away.* The light from this galaxy traveled for 12.9 billion years in order to reach us, so we are seeing this galaxy as it was 12.9 billion years ago, or about 800 million years after the Big Bang. The very first galaxies are thought to have been quite small, and these merged into mid-sized galaxies. With the best telescopes, astronomers are now studying many of these "second generation" galaxies that are about 12 billion light-years

* The oldest stars yet seen in our galaxy and in a few other galaxies are some of the first stars that formed about 400-500 million years after the Big Bang. This suggests that galaxies started to form shortly after the first stars were born, and that our own galaxy is one of the oldest in the universe.

away, offering a direct view of the universe when it was one to two billion years old. Because they have the use of powerful telescopes, astronomers are the only historians who can actually *see* the past!

As we look at galaxies from the first 6 billion years of the universe, a trend becomes apparent. In the first few billion years there were many small, disorganized galaxies (astronomers call these *peculiar* galaxies) and no spiral or elliptical galaxies, because these take much longer to form. As time passed, there were more and more spirals and ellipticals, and fewer small new galaxies. Finally, after about 6 billion years, galaxy formation virtually stopped. In other words, the first six billion years in the life of the universe was a time of prolific galaxy formation and mergers into larger galaxies, but apparently this work was finished as the universe moved into the second half of its life.

The cessation of galaxy formation coincided with something else very strange and surprising that was first discovered in 1998 by two teams, one based at Harvard University and headed by Adam Riess and Brian Schmidt, and the other based at the University of California Berkeley and headed by Saul Perlmutter. They found that the expansion rate of the universe has apparently been *increasing* since about halfway through its life (Reiss et al. 1998; Overbye 2003). Reiss, Schmidt, and Perlmutter shared the 2011 Nobel Prize in Physics for this discovery. Before this was observed, cosmologists suspected that the expansion rate was *decreasing* because of gravity pulling the universe together, but they could not determine the rate of decrease. Now it appears that at the very largest scales, something is pushing the universe apart. No one yet knows exactly what this outward push is, but it has come to be called *dark energy*. Dark energy is even more elusive than dark matter because it cannot be detected in any way yet known—but it apparently makes up about three-fourths of the contents of the universe (see Figure 2-4). It does not clump together like matter, but permeates all of space smoothly and exerts an outward push on everything.

Figure 2-4

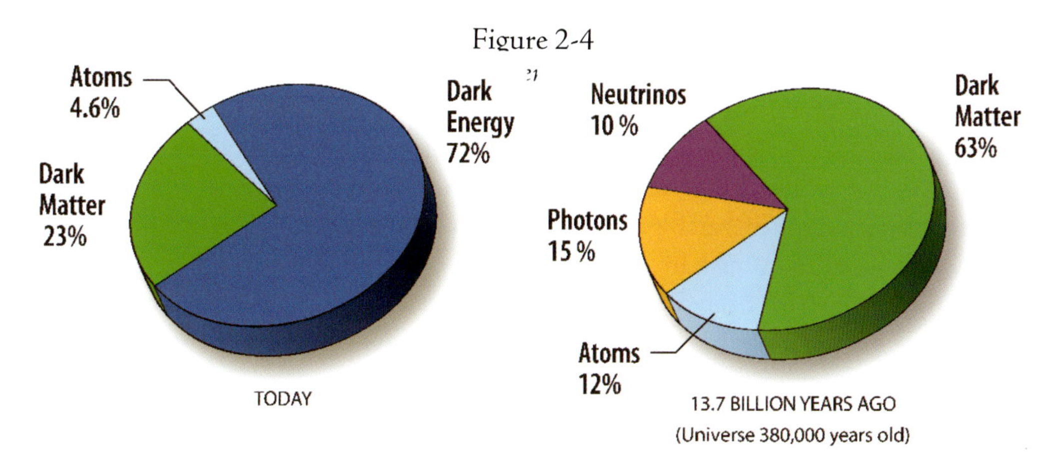

NASA/WMAP Science Team

During the first half of the life of the universe, matter was more closely packed together because the Universe was smaller, so the inward pull of gravity dominated over the outward push of dark energy. But as the universe grew and the average density of matter decreased, dark energy overtook gravity and began to push things apart. The crossover point, when dark energy began to dominate over gravity, seems to coincide with the significant drop in galaxy formation about 6 billion years ago.

Although we do not yet know what dark energy actually is, it may be just what Einstein inserted into his equations in 1917 as he modeled the universe using his newly developed General Theory of Relativity. This *cosmological constant* was a repulsive factor intended to counteract the effects of gravity and thus give a static universe, but it was deemed unnecessary when Hubble showed in the late 1920s that the universe was not static at all, but expanding. Until his death in 1955, Einstein regretted the "blunder" of the cosmological constant, but with the discovery in 1998 that the expansion rate of the universe has been accelerating since about halfway through its life, a repulsive factor is needed once again. The cosmological constant has now been resurrected in the form of dark energy, and some cosmologists now suspect that dark energy is the same as "vacuum energy," the mysterious force that blew the universe apart during the inflationary period, and the "repulsive gravity" that the theory of loop quantum gravity predicts during the moments just after the Big Bang.

Although we still know very little about dark energy, the one thing that seems certain is that it is guiding the overall fate of the universe, and that fate seems to be rather unexciting. It now appears that the universe will simply fizzle out by expanding forever. If this is true, the distant galaxies will eventually disappear from our sight as the universe slips into cold, empty oblivion. We may be living in one of the most exciting eras in the history of the universe, when many galaxies are still visible, star formation is still prolific, and life thrives on at least one planet.

The life of the universe, starting about 400 million years after the Big Bang, has been a continual process of star formation and galaxy formation, although these processes have slowed down considerably in more recent times. Galaxies are among the oldest things in the universe; they evolve and grow continuously through the mergers of small galaxies into larger ones. Even our own Milky Way appears destined to eventually merge with our sister, the Andromeda Galaxy. But in contrast, there have been many generations of stars. Although some stars (small ones) have been here from the beginning, 400 million years after the Big Bang, many others (large ones) are long since dead, their material recycled into newer generations of stars. It is stars that play the central role in making life possible in the universe, so we must now turn our attention to stars, to understand what they are and how they work.

Figure 2-5
Main Events in the Later Life of the Universe

Name	Time after Big Bang	What was happening
Nucleosynthesis Epoch	1 sec – 3 minutes	Protons and neutrons combined to make nuclei of deuterium, helium, and lithium. Dark matter began forming the cosmic web.
Recombination	380,000 years	The Universe cooled to about 3000 degrees, allowing electrons to attach to nuclei, forming the first neutral atoms. Photons of light now escaped into space. These are seen today as the *cosmic microwave background*.
Dark Matter Epoch	< 400 million years	It was still too hot for gravity to form stars, but dark matter was invisibly clumping into large filaments and "halo" structures known as the cosmic web that would become the framework for galaxies and clusters of galaxies.
Star Epoch begins	400 million years	The first stars formed, producing ultraviolet radiation that re-ionized clouds of hydrogen. Stars begin producing heavier elements.
Galaxy Epoch begins	600 million years	The first galaxies began to form in halos of dark matter. Smaller ones merged into larger ones, some forming massive spiral galaxies like the Milky Way.
Dark Energy Epoch	> 6 billion years	The mass density of the universe fell enough so that the outward push of dark energy began to dominate over gravity at the very large scale. The expansion rate of the Universe now began to increase, while galaxy and star formation declined.
Sun Epoch begins	9.0 billion years	The Sun formed.
Earth Epoch begins	9.1 billion years	The Solar System and Earth formed.
The present	13.75 billion years	Here we are.

Figure 2-6
Evolution of the universe

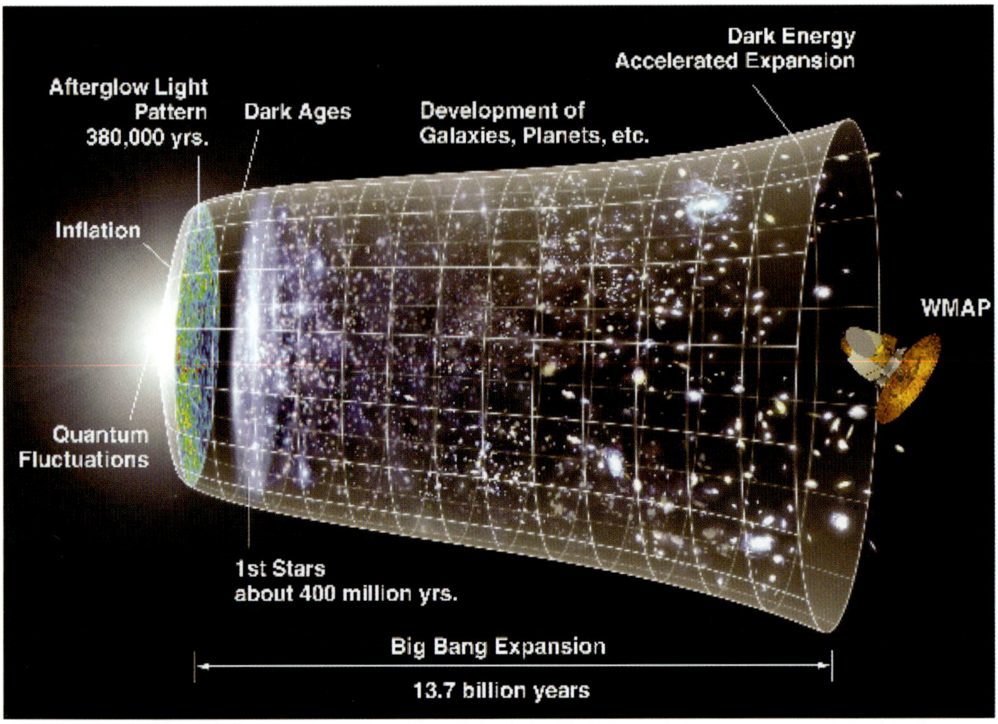

Courtesy: NASA/WMAP Science Team

The Physics of Stars

When we think of "nature," most of us probably imagine things like birds, bees, flowers, trees, oceans, and mountains – things that are familiar to us here on Earth. But these things are extremely rare in the universe, and not at all typical of nature. If you had to pick one thing that really typified nature, something that summed up nature more than anything else, a very good candidate would be stars. We now know that there are at least 10^{22} stars in the universe—far more than *the number of grains of sand on all the beaches of Earth!* And new stars are being born all the time, as some also die in massive supernova explosions.

Until the 1930s the nature of stars was a mystery. In order to be seen, even the closest stars to the Earth have to produce enormous amounts of energy per second. The most energetic processes known in the early decades of the1900s were chemical reactions such as the burning of coal or the explosion of gunpowder. But even the most energetic of these cannot come close to producing the brightness of stars. With the discovery of the neutron in 1932, the world of the atomic nucleus opened up, and the field of nuclear physics was born. The nuclear reaction can produce a million times more energy than the chemical reaction, and so the mystery was solved: stars are nuclear-powered.

Stars are the workhorses of the universe, making planetary systems as they form, providing abundant energy for those planets, and manufacturing elements in their core as they produce energy. All of the atoms heavier than hydrogen and helium that make up our bodies, and Earth, were manufactured in ancient stars, now long gone. The first generation of stars that formed about 400 million years after the Big Bang contained only hydrogen and helium when they were born, because the heavier elements did not yet exist. By contrast, the Sun formed much later and contains all of the first 92 elements (up to uranium). Because the Earth formed from the same material as the Sun, it too contains all 92 elements.

All stars begin their life in the same way and go through the same early stages of life, but their later lives and their final fates are radically different, depending on just one thing: their starting mass. Stars can have starting masses as small as 10% of the Sun's mass or as large as 100 Solar-masses, but anything outside this range cannot become a star (Cowen 2010). Stars like the Sun, or a little smaller, make up about 95% of all the stars we see, while larger stars are rare. There are two reasons for this rarity: first of all, star formation is heavily weighted toward lower mass stars; the second reason is that large stars have very short lives, so they are not around for very long (see Figure 2-7). Their lives are so short because of their enormous brightness (luminosity): to produce so much energy per second they must consume their hydrogen at a very high rate. They literally use themselves up in a very short time.

Figure 2-7
The Length of a Star's Life

Mass of Star compared to Sun	Length of Life in years ($t \alpha 1/M^{2.5}$)	Luminosity compared to Sun ($L \alpha M^{3.5}$)
0.1	3.2 trillion	0.0003
0.3	205 billion	0.015
0.5	57 billion	0.1
0.8	17 billion	0.5
1.0 (like the Sun)	10 billion	1.0
2.0	1.8 billion	11
4.0	0.3 billion = 300 million	130
10	30 million	3200
40	1 million	405,000

Stars are born in vast nebular regions of gas and dust like the Horsehead Nebula (Figure 2-8). The first stars probably formed as gravity pulled cooler, denser material into a blob of dark matter. Today dark matter is too diffuse to drive star formation; instead, clouds of matter radiate energy and cool, and then collapse under the effect of gravity. As material falls inward, it compresses, heats up, and the whole cloud begins to slowly rotate. This is a *protostar,* but not yet a true star. As gravity further compresses the protostar, it gets hotter and rotates faster. If it is massive enough to begin with, and can heat up to about 15 million degrees, *nuclear fusion* begins in the core. It is the advent of nuclear fusion that marks the birth of the star, and it is fusion that provides the vast energies that pour out from stars every second. Hans Bethe first worked out the details of nuclear fusion in stars in 1938, solving the longstanding mystery of their energy source, and received the 1967 Nobel Prize in Physics for this and other pioneering work in nuclear physics. For more details about fusion, see *Exploring Deeper* on the next page about nuclear fusion.

Figure 2-8
A Hubble Space Telescope photo of the Horsehead Nebula, a "star nursery" where new stars are being born. It is about 1500 light-years away from earth (well within our galaxy) and about 3.5 light-years wide.

Courtesy: Canada-France-Hawaii Telescope/Coelum, J.-C. Cuillandre & G. Anselmi

When a protostar becomes a star—that is, when the core temperature gets high enough for hydrogen fusion to begin—a balance is reached between the inward pull of gravity and the outward push of radiation. In this stable state, the star now shines steadily for a long time, until all the hydrogen in the core is turned into helium. Stars in this "hydrogen burning" stage of life are called *main sequence stars* and this stage fills the largest part of the life of any star. The Sun will spend about 10 billion years in this stage and it is presently about halfway through this part of its life. As Figure 2-7 shows, massive stars have huge luminosities and much shorter lives.

Exploring Deeper

What is Nuclear Fusion?

In the simplest terms, nuclear fusion is the joining together of two light nuclei to form a heavier one, most commonly turning hydrogen into helium. Figure 2-9 depicts the fusion process that is occurring in the cores of stars the size of the Sun. Normally, two protons do not want to be near each other because they are electrically positive and repel each other strongly. But if they can somehow get very, very close to each other, they will suddenly slam together because of a much stronger attractive force, the *strong nuclear force*. Protons can get this close and fuse together if they are at a high enough temperature. When protons reach a temperature of about 15 million degrees they are moving so fast that they can fuse together as they collide head-on.

Figure 2-9

The fusion of hydrogen into helium that occurs inside the Sun.

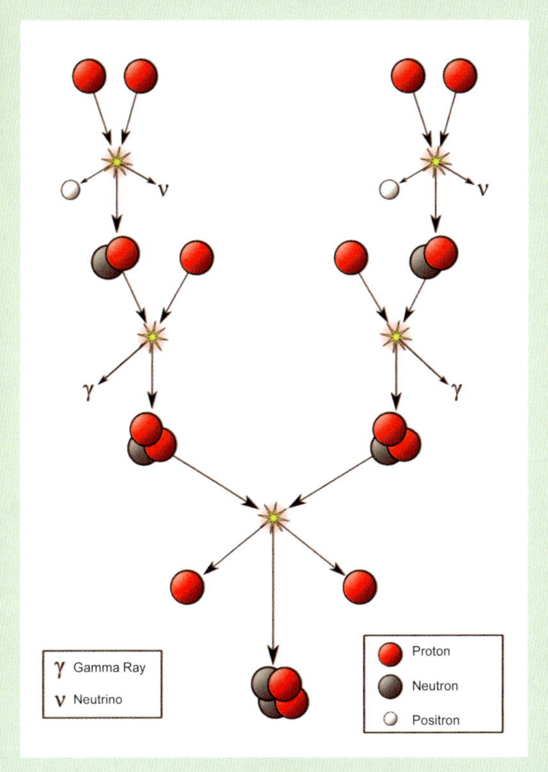

Wikimedia Commons

Nuclear fusion does two important things: it creates heavier elements and it releases large amounts of energy. At a temperature of 15 million degrees, hydrogen fuses into helium, but this is the *coolest* type of fusion. At about 100 million degrees, helium fuses into carbon, and at high enough temperatures (billions of degrees) all of the first 26 elements, up to iron, can be created. As far as we know all of these types of fusion reactions can happen only inside stars, because this is the only place we know that can be this hot.

The energy produced by fusion is accompanied by a *disappearance* of mass. Einstein's famous equation $E = mc^2$ says just this: a tiny amount of mass can disappear and turn into a large amount of energy. In the fusion process, this huge energy is released mostly in the form of gamma photons.* In the core of the Sun where hydrogen is fusing into helium, mass is disappearing at a rate of about 5 *million tons per second*, producing about 10^{26} watts of energy at the surface. But even at this staggering rate the Sun will not use up its mass any time soon—its 10^{30} kilograms of mass is so large that it can keep this up for billions of years. The gamma photons produced by the fusion in the core of the Sun take a very long time to reach the surface because they get absorbed by the plasma particles of the Sun and re-emitted at longer wavelengths to eventually reach the surface of the Sun as the visible light that we see streaming out to us. Our best models of conditions inside the Sun show that the energy produced by nuclear fusion in the core takes millions of years to reach the surface of the Sun!

Can humans achieve nuclear fusion here on Earth and harness its energy? That is, can we create the conditions found in the core of a star? The answer is yes and no. In 1952, American scientists achieved the first fusion reaction, but it was an *uncontrolled* reaction known as the *hydrogen bomb*. This bomb was a big step beyond the first nuclear weapons unleashed in 1945 at Hiroshima and Nagasaki—these smaller bombs are referred to as *atomic bombs* and are based on the *nuclear fission* of uranium or plutonium, rather than the *nuclear fusion* of hydrogen. Hydrogen bombs use the smaller atomic bombs as detonators to create the 15 million-degree-temperature necessary to ignite fusion. Unfortunately, the vast energy can only be released in one burst and is useful only for destructive purposes. The problem of achieving *controlled* fusion is still unsolved, though scientists have worked on it for fifty years. They simply have not been able to bring a large quantity of hydrogen up to a temperature of 15 million degrees in a controlled fashion. If and when fusion reactors become a reality, a vast new source of energy will become available with relatively harmless waste products.

The main part of the life of any star is over when all of the hydrogen in the core has been fused into helium. At this point, hydrogen fusion continues in a shell surrounding the core, while the core collapses and heats up. When the core temperature reaches about 100 million degrees, helium fusion begins, and this entire process causes the star to expand dramatically, becoming a *red giant*. When the Sun reaches the red giant stage, it will be so large that it will engulf and vaporize Mercury and Venus, while Earth and Mars will be scorched into rocky cinders. But this is not something to worry about because it will not happen for another 5 billion years—a bit longer than Earth has already existed.

A star like the Sun remains in the red giant stage, fusing helium into carbon and oxygen, for about 20% as long as it spent in the first phase of its life. But how a star ends its life varies widely, depending on its starting mass. For smaller stars, up to

* Some energy is also in the form of the kinetic energy of the fusion byproducts.

about 4 times the Sun's mass, helium fusion in the core ends, but helium and hydrogen fusion continue in surrounding shells. Eventually the star sheds its outer layers, ejecting as much as 50% of its mass. This mass, rich in the helium, carbon, and oxygen created in the star, is hurled out into space to form a *planetary nebula* (a misnomer since it has nothing to do with planets). The remaining carbon-oxygen core can never get hot enough for fusion to continue and it cools down to end its life as a white dwarf that eventually becomes a cold black dwarf. The Sun will eventually fade into obscurity as a white dwarf, and finally a black dwarf about the size of Earth.

Large-mass blue stars end their lives much more dramatically. As with smaller stars, they spend most of their lives fusing hydrogen into helium. Then, in a series of stages, each shorter and hotter than the last, fusion continues, creating successively heavier elements. But nuclear fusion, even at billions of degrees, can only create the first 26 elements, up to iron.* The core eventually fills with iron and cools, then collapses and bounces back suddenly. The mass and gravity of the star are so large that the collapse and bounce result in a stupendous detonation—a *supernova*. The star explodes in one of the most energetic and spectacular events known, hurling all of its matter, made up of the first 26 elements, out into space for billions of miles.

Figure 2-10
The Life Stages of Stars

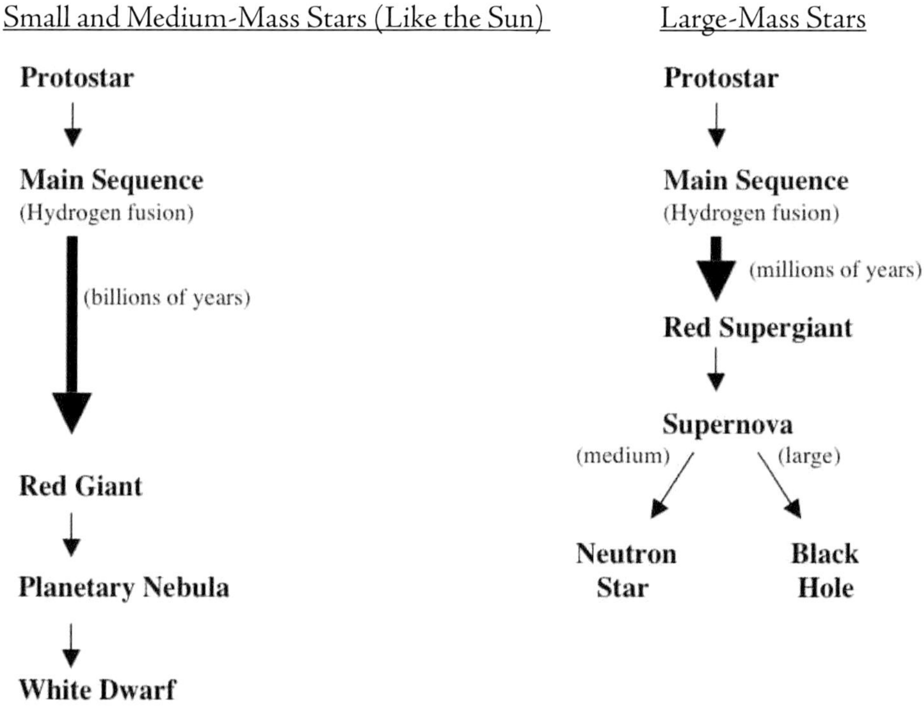

Small and Medium-Mass Stars (Like the Sun)

Protostar
↓
Main Sequence
(Hydrogen fusion)
↓ (billions of years)
Red Giant
↓
Planetary Nebula
↓
White Dwarf

Large-Mass Stars

Protostar
↓
Main Sequence
(Hydrogen fusion)
↓ (millions of years)
Red Supergiant
↓
Supernova
(medium) ↙ ↘ (large)
Neutron Star **Black Hole**

* The *s process* (for slow neutron) that occurs in the envelopes of red giants can produce elements heavier than iron, up to lead.

But the explosive force of the supernova itself creates even heavier elements, up to uranium and beyond. It is the supernova process that seeds the universe with the elements heavier than iron. The remnants of supernovae, rich in heavy elements, can later condense into new stars that will contain heavy elements (these are called "metal-rich" stars). Because the Sun and Solar System are rich in heavy elements, we can be certain that a supernova was involved in some way in its formation (more in the next section).

A massive supernova explosion leaves something behind: the star's core is crushed to an unimaginable density, with protons and electrons merging together into neutrons. A tiny *neutron star* remains, and is so compact that a teaspoonful of its neutron matter would weigh five million tons. It spins rapidly and projects a beam of radiation that can be detected from Earth as pulsating signal. These pulsating neutron stars were named *pulsars* when they were first discovered in the 1960s.

Figure 2-11
Stages in the Brief Life of a 25 Solar-Mass Star

Stage	Length of Stage	Core Temperature
Hydrogen fusion	7,000,000 years	40 million °K
Helium fusion	700,000 years	200 million °K
Carbon Fusion	600 years	600 million °K
Neon fusion	1 year	1.2 billion °K
Oxygen fusion	6 months	1.5 billion °K
Silicon fusion	1 day	2.7 billion °K
Core collapse	0.25 seconds	5.4 billion °K
Core bounce	thousandths of a second	23 billion °K
Supernova	10 seconds	

But an even stranger fate awaits the most massive stars. When the supernova occurs, the core collapse is so extreme that nothing can stop it. Its gravity is so strong that space and time around it are completely warped and a hole in the fabric of space and time is created. Nothing can escape from it—not even light. This is a *black hole*.

Exploring Deeper

Black Holes and Quasars

The possibility of a "dark star" from which light could not escape is not a new idea. This was first suggested by John Michell in 1784, and independently by Pierre-Simon Laplace in 1796. But no one paid much attention to the bizarre idea until 1915, when Karl Schwartzschild showed that Einstein's newly published *General Theory of Relativity* predicted such a thing as the final stage in the life of a very massive star. For the remainder of the twentieth century, physicists and astronomers including Arthur Eddington, Subrahmanyan Chandrasekhar, Robert Oppenheimer, David Finkelstein, Roy Kerr, John Wheeler, Roger Penrose, and Stephen Hawking continued to work out the details of black holes. All of this theoretical work pointed unquestionably to their existence, but no one could actually *see* a black hole, because they emit no light. Finally, in the 1980s, indirect evidence accumulated for the reality of black holes, because of the effects of their gravity. In some cases a companion star can be seen that appears to wobble as it orbits the invisible black hole; in other cases x-rays can be detected that are the result of matter falling into a black hole; and most recently, gravitational lensing—the bending of light as it passes through a strong gravitational field—has been used to locate black holes.

Soon after John Wheeler coined the term "black hole" in 1967, the idea became mainstream and captured the imagination of the general public. Scientists and science fiction writers have speculated about what lies inside a black hole and what would happen to someone if they fell into a black hole. Some scientists have speculated that when mass disappears down a black hole, it might spew out somewhere else as a *white hole,* perhaps in another universe. Or, a black hole might be a type of tunnel through space-time—a *worm hole*—that comes out in some other part of our universe. A worm hole might some day provide a "short cut" that would allow long-distance space travel in much the same way a tunnel through a mountain range cuts off distance and saves time.

It has now been confirmed that very large black holes, called *supermassive black holes,* reside at the centers of galaxies, including our own. Our own galactic black hole has a mass of 4 million Suns and emits a wide spectrum of radiation as it swallows up nearby mass. But it is relatively small and quiet compared to others that have been detected. In 1978 Peter Young, from Cal Tech, showed that the motions of stars in the galaxy M87 implied the existence of a central mass of several billion solar masses, though firm evidence for this would not come for another twenty years. Today, highly active *ultramassive black holes* with masses of around 10 billion Suns are well accepted (Petit 2008).

The existence of ultramassive black holes also solves a longstanding mystery. Beginning in the 1960s astronomers found a new class of objects that emitted intense radio waves and had the largest redshifts yet measured. Such high redshifts indicated that these objects were very far away—billions of light-years—making them the most distant objects yet seen; but they would have had to be extremely bright in order to be detected at such great distances. Calculations showed that these quasi-stellar objects, or *quasars,* were emitting as much energy as an entire galaxy, yet they seemed to be more like the size of a large star. There was no energy source known at the time that could account for how quasars could be so small and so bright at the same time. Through the next decades, many more quasars were cataloged at distances extending out to 10 billion light-years, but never closer than about 1 billion light-years. Whatever they were, they apparently existed in the earlier days of the universe, but not in the last billion years.

Many theories were advanced to explain quasars, but not until the final years of the 20th century did the evidence begin to suggest that they were ultramassive black holes being fed enormous quantities of gas. We now know that the ultramassive black holes at the hearts of the largest and most active galaxies are swallowing up prodigious quantities of mass—sometimes the equivalent of 60 Suns each year—and ejecting powerful jets of matter. In these processes, intense radiation is emitted, including radio waves, x-rays, and gamma rays. The black holes at the centers of our own galaxy and our neighbor Andromeda are not large enough, nor is there a sufficient supply of gas to feed them, to shine as brightly as a quasar. Because we see quasars only at very large distances, it must be that they existed in the earlier days of the universe, when galaxies were more closely packed with matter and galaxy mergers were common. In more recent times, the universe has quieted down, and there are apparently no more quasars.

Figure 2-12

Close-up of a star.

This photo, taken in ultraviolet light, captures a quiet day on the Sun. The plasma streamers erupting from the bright area are 60,000 miles tall (almost eight Earths). The brightest areas are at a temperature of a million degrees and the dark areas are at about 10,000 degrees. This part of the Sun, the chromosphere, is hotter than the photosphere below it which is the "surface" of the Sun that we see shining at a cool 6000 degrees Kelvin.

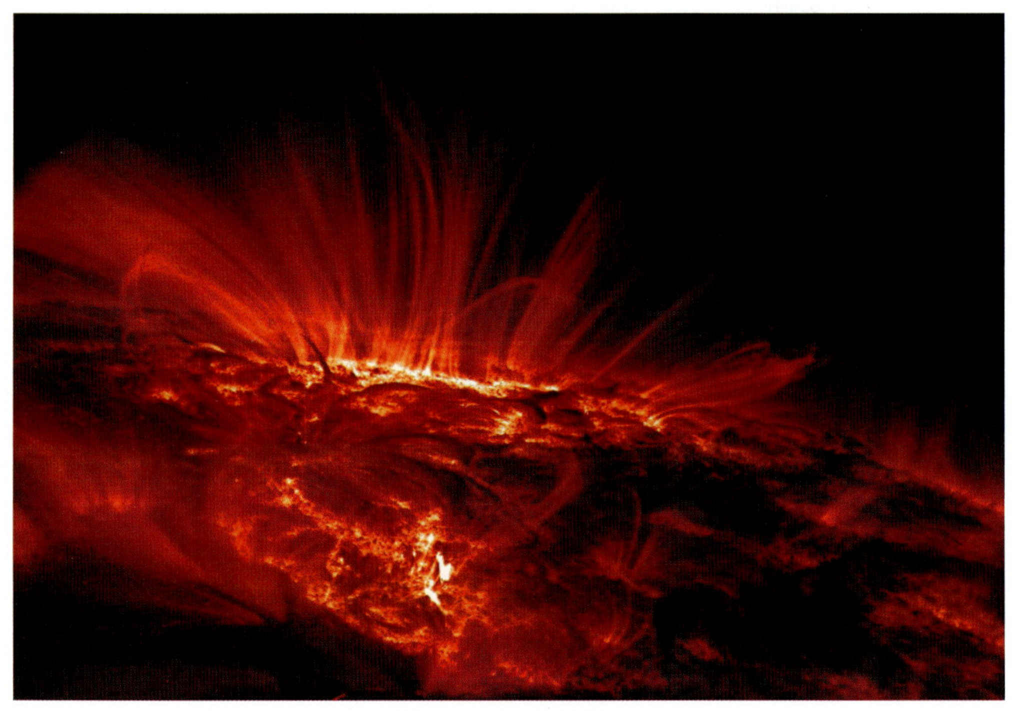

Trace project/NASA

Astronomers understand stars so well because they have studied large numbers of them, each in a different stage of life. They can watch stars being born in giant nebular regions, and small stars dying slowly as white dwarfs, and occasionally they see massive stars die suddenly in supernovae. This is not so different from learning about the life stages of humans by observing a large number of people, each at a different age. We would not have to wait an entire human lifetime in order to piece together what the life stages were like; and likewise with stars, we do not have to watch for billions of years to discover what the lives of stars are like.

The first stars that formed about 400 million years after the Big Bang were composed only of hydrogen and helium because the heavier elements did not yet exist. These first-generation stars began to manufacture the heavier elements, but they were massive and had short lives, dying quickly in supernova explosions that hurled new matter rich in heavy elements across space. In the vast regions of gas and dust that remained, new stars were born, and these new-generation stars now contained the heavier elements like carbon, oxygen, and silicon. Many generations of large stars have lived and died since the first stars began to shine and these have continuously churned out the ingredients for stars like our own Sun with all 92 elements. In succeeding generations, smaller stars became abundant, and some of these, with extremely long lifetimes, are still shining today.

The very large blue stars are important because they create the heavy elements and cast these out into space to later become new generations of stars that are rich in heavy elements. But do the smaller stars, like our own yellow star that will eventually fade into obscurity as a white dwarf, play some equally important role in the cosmos? Indeed they do, because Sun-like stars provide the steady, long-term energy source for the planetary systems that form around them. And given enough time and the right conditions, some very special planets might possibly spawn and nurture life.

Chapter 3
Sun and Earth

The Birth of the Sun

Until recently, most astrophysicists believed that the Sun formed alone from a cloud of dust and gas that contracted to become our star and the surrounding Solar System. This view was based mostly on looking at our own secluded location in a quiet neighborhood of the galaxy, where stars are typically separated by about four light-years. The sky at night looks black because nearby stars are so far away: our most sophisticated spacecraft today would take about 80,000 years to reach the nearest star (and this is why there is no serious talk of sending a spacecraft to a nearby star at any time in the foreseeable future).

But in recent years astronomers have been able to study many star clusters within our galaxy and in other galaxies. These are huge collections of stars containing thousands or sometimes millions of stars packed into a small volume. Star clusters have been seen since 1609 when Galileo first saw the Beehive Cluster, a small nearby cluster that is now known to contain about 350 stars. Recent studies of star clusters show that the stars in a particular cluster are, strikingly, all about the same age. This suggests that they were all born together, at roughly the same time. With powerful instruments like the Hubble space telescope, or the 10-meter Keck telescope on top of Mauna Kea in Hawaii, astronomers have now studied hundreds of clusters in detail and concluded that they are cocoons of sorts where stars are born. Instead of being born alone, it now seems that stars are typically born with thousands of siblings.

Observations of young star clusters show that the first stars to form are large blue stars with masses of 15 to 25 times the mass of the Sun. Later a full range of smaller stars form, and the smaller they are, the more abundant they are, down to the tiniest with a mass of about a tenth the mass of the Sun. Astronomers estimate that for each large star, there will be about 1500 small stars, all packed into a few light-years. But the large stars live very short lives and soon die in spectacular supernova explosions and when they do, any nearby material will be seeded with heavy elements.

It is just such a scenario that apparently led to the birth and infancy of the Sun. Astronomer Simon Portegeis Zwart from Leiden University in the Netherlands has combined theoretical models with observations to work out many of the details of the Sun's birth (Portegies Zwart 2009). According to Zwart's theory, it all started roughly 5 billion years ago in a large cloud of gas and dust that was located in the Orion Arm of the Milky Way Galaxy, about 33,000 light-years from the center of the galaxy and

200 light-years above the galactic plane. This birthplace was about 3000 light-years away from our present location (we are now about 30,000 light-years from the center of the galaxy). This cloud gave birth first to one large star, and over the next few million years the smaller stars condensed out. Our birth cluster probably contained some 1500 to 3000 stars, all packed within the space of about five to ten light-years. One of these was the Sun, and the rest were the Sun's siblings.

When star clusters are too closely packed, more stars will end up orbiting each other as double or triple stars, and planet formation around individual stars is impaired by the chaotic gravity of other nearby stars. Because our cluster was not too tightly packed with stars, our star was able to survive as a single star and form a magnificent and stable planetary system. Recent observations of other star systems are beginning to suggest that such "perfect" star systems may be rare.

Sometime just after the Sun and Solar System had formed, the original large blue star that resided in our cluster reached the end of its life and exploded in a supernova. Many nearby stars were "polluted" with the full range of heavy elements that a supernova produces. This supernova was probably very close to the Solar System—perhaps as close as 0.07 light-years—and injected the Sun and Solar System with a rich supply of the 92 elements that would make life on Earth possible.

This scenario, in which the original large star went supernova *after* the Solar System had formed, received support when two primitive meteorites were analyzed in 2003 and found to contain a rare isotope called Nickel 60 (Tachibana and Huss 2003; Huss, and Tachibana, et al. 2008). This isotope exists only as a result of the radioactive decay of Iron 60. But Iron 60 is not found naturally in the Solar System, and must have been synthesized in a supernova. It appears that our nearby supernova must have injected Iron 60 into the early Solar System after it had formed. Some of this material containing Iron 60 was preserved as meteorites, and within these meteorites the Iron 60 decayed into Nickel 60, which remained locked inside. The presence of Nickel 60 is the fingerprint of the supernova. After the supernova occurred, the density of our star cluster was reduced, weakening gravity, and the cluster drifted apart. Each of our sibling stars went its separate way, and none of them are close to us today.

Another close encounter with a sibling may explain a long-standing puzzle about our Solar System. While the orbiting planets all have nearly circular orbits lying roughly in one plane, the comets in the outer solar system have highly skewed orbits that are far off the main plane. No theory of planetary systems has yet been able to explain why. But if the Sun was born in a cluster with many siblings, one of them may have passed very near the Solar System, stirring up the comets. Such a passerby must not have come close enough to disrupt the planets. Theoretical calculations suggest that this star passed by somewhere between 100 AU and 1000 AU from the Sun (one Astronomical Unit, or AU, is the Sun-Earth distance—Pluto is about 40 AU from the Sun).

Where Do Planets Come From?

Until recently our ideas about how planetary systems form came only from looking at our own Solar System. But in 1988, astronomers reported the first tentative evidence of *extrasolar planets (exoplanets)*—planets that circle *other* stars (not our Sun) (Campbell et al. 1988). By 1995 a planet orbiting a Sun-like star was confirmed (Mayer and Queloz 1995), and since then planet-hunting astronomers have refined techniques and instruments to discover many more exoplanets. By 2011 nearly 600 had been cataloged* and most of these were large Jupiter-like planets with enough mass to make their star wobble measurably. But starting in 2010 the potential for discovery increased dramatically as the Kepler spacecraft came online. Launched in March of 2009, Kepler carries a super-sensitive telescope and photometer that looks at 156,000 nearby stars during its three-year operation time that began in early 2010. Its photometer can detect the minute drop in the brightness of a star caused by a planet passing in front of it (a *transiting* planet). At the end of its first year of operation Kepler scientists announced 1,235 new candidate planets, including 54 where life might be possible. The Kepler mission should go far to determine the likelihood of other Earth-like planets that could harbor intelligent life, but for the time being we know of only one such planet (see *Science and Discovery: Are We Alone?* on pg. 49).

With these new capabilities emerging since about 1995, astronomers have begun to study other planetary systems that are in the process of forming and a wide variety of mature systems. Planetary scientists once thought that the formation of planetary systems was fairly orderly and predictable, but they now suspect that it is a very chaotic and violent process. There is little doubt that virtually all stars form some kind of a planetary system around themselves, though many of the planets probably do not survive. In a real sense, planets are a part of the star—they form in the same process, and from the same material. Because there are so many stars, planets must also be very abundant, yet we do not know how common *Earth-like* planets are. Many scientists suspect that life could begin in numerous places, but whether it could evolve in safety for billions of years, as it did on Earth, to become complex and intelligent, is an open question.

Based on observations of planetary systems around other stars, scientists have theorized that planetary systems form in a series of stages, though some of these stages may have different outcomes. There are still alternative theories of planetary formation and new ones being proposed (Cowan 2011). Below is a description of the stages of formation of our own Solar System, based on the predominant theory (Lin 2008), a process that took about 100 million years:

---

* Jean Schneider of the Paris Observatory keeps a current (updated daily) database of all known extrasolar planets on *The Extrasolar Planets Encyclopedia* at http://exoplanet.eu/

Stage One: The Star. An interstellar cloud of gas and dust collapsed to form the Sun, surrounded by a flat disk of material revolving around it. After a few hundred thousand years of gravitational collapse and heating, nuclear fusion began and the Sun became a star.

Stage Two: Planetesimals. Tiny particles of dust and ice in the surrounding disk began to collide and stick together, slowly growing into *planetesimals* about one kilometer in size. During this stage something else very important happened in the Solar System – a "snow line" formed. In the inner part of the young system, where it was hot, all the water was vaporized and driven, along with hydrogen and helium gas, to the outer regions of the system. Outside this snow line, water existed as ice; this line later divided the Solar System into an inner region with rocky bodies (Mercury, Venus, Earth, Mars, and the asteroids), and an outer region filled with icy bodies (Jupiter, Saturn, Uranus, Neptune, and the comets).

Stage Three: Planetary embryos. Over the next few million years, planetesimals continued to collide, sometimes breaking apart and sometimes sticking together. As some grew in mass, their gravity also grew and they could attract even more mass. Eventually the Solar System became filled with thousands of planetary embryos about the size of the Moon, all orbiting the Sun. These continued to collide violently, some growing bigger and some breaking apart.

Stage Four: The first gas giant. A planetary embryo orbiting just outside the snow line probably grew to about ten times the mass of Earth and began to attract some of the hydrogen and helium that was pervasive in the outer Solar System. The infalling gas heated up, causing some to be driven off, but somehow a balance was struck between heating and cooling so that the mass grew into a gas giant with abundant hydrogen and helium. There may have been many false starts before this tricky process finally yielded Jupiter.

Stage Five: More gas giants. Once Jupiter formed, its vast gravity helped other gas giants form more easily. Next came Saturn, which was much like Jupiter, then Uranus and Neptune, which were mostly ice. Many of the remaining icy planetesimals were flung outward to the far reaches of the Solar System to become the *Kuiper Belt* and the *Oort Cloud*. When icy planetesimals from these regions fall inward toward the Sun, they become comets.

Stage Six: The rocky inner planets. Inside the snow line, the early Solar System was still filled with hundreds of planetesimals and embryos made of high-boiling point materials like iron and silicates – the stuff of rocks. The water and other volatiles had long ago boiled off and been driven to the outer Solar System. Probably disturbed by the gravity of the outer gas giants, these inner bodies now battled it out in a violent free-for-all of collisions. The survivors became the asteroid belt,

Mars, Earth, Venus, and Mercury. The asteroid belt was probably a planet that could never coalesce into one mass because it was continually pulled apart by the competing gravities of the Sun and nearby Jupiter. In one final gigantic collision, a Mars-sized protoplanet struck the young Earth and threw out a large amount of debris that coagulated into the Moon. Moon rocks brought back by astronauts have been dated to about 4.53 billion years ago and this is now widely accepted as the approximate age of the Earth-Moon system. The major bodies of the Solar System were now finished, though it would be another 500 million years or so before all of the remaining debris around the Solar System was "mopped up."

Compared to the hundreds of other planetary systems we have recently studied, our Solar System seems to be a model of simplicity and stability. Our star is long-lived—it has shined for about 5 billion years and has another 5 billion to go before it becomes a red giant—and it is a single star, which only about half of all stars are. Double or multiple stars have such chaotic gravitational effects that stable planetary systems are unlikely. But the planets in our system orbit in nearly perfect circles and are spaced ideally to prevent gravitational disturbances that could tear the system apart. Gravitational models show that our system is a full house—the insertion of just one more Earth-mass planet inside the snow line would destabilize the whole thing.

It might appear that our model Solar System is the result of some grand design, but the latest studies of other planetary systems suggest that it is simply what survived after a hundred million years of chaotic collisions and dynamic evolution. Yet there can be no doubt that our planet, the third one out from the Sun, is rare. Our large moon is as big as Mercury and is really a sister planet that has protected us from meteor bombardment. This double planet, Earth-Moon, also formed at the right distance from the Sun for liquid water to exist—a key requirement for life. Every star has a doughnut-like band around it where these ideal conditions exist—a *habitable zone*—and Earth is the one planet in our system that squarely occupies the habitable zone for our star. But even with this ideal and lucky set of conditions, many more things still had to fall into place on the young Earth 4.5 billion years ago, before life was possible.

Science and Discovery

Are We Alone?

In 1950 the colorful and irreverent physicist Enrico Fermi posed a now-famous question over lunch with some friends: Where is everybody? He was talking about the other forms of intelligent life in the universe who were so conspicuously absent. Why had we seen no signs of any kind that there was life elsewhere in the universe? Astronomers knew by the mid-twentieth century that the universe was filled with an enormous number of stars (about10^{22} or so) and they suspected something that we have now confirmed: that most stars have a planetary system of some kind around them. Given the staggeringly large number of planets that must be in the universe, the odds seem very high that many of these would be Earth-like; and that on some of these Earth-like planets, life could begin and be nurtured for billions of years so that it could evolve into intelligent beings with technology, communications, and perhaps even spaceships.

In just our own galaxy there are several hundred billion stars, and if just one star in a million harbored an Earth, there would be several hundred thousand planets where intelligent life could arise; and if just one in a thousand of these possibilities panned out, there would still be a hundred or so advanced civilizations capable of contacting us. Science fiction writers have long envisioned visits from alien neighbors, both malevolent and friendly, yet today there is not one shred of evidence to confirm the existence of any form of life beyond Earth. If we have any neighbors out there, they seem to be *very* quiet. This puzzling disconnect between what we would expect and what we have actually seen has been called the Fermi Paradox. Since the day Fermi posed the question, many possible explanations have been offered,* including:

- We ARE alone in the galaxy and maybe even in the universe.
- Intelligent life tends to destroy itself before it can make contact with other intelligent life. Humans certainly display this tendency.
- Life is periodically destroyed by natural extinction events such as asteroid impacts or extreme changes in climate.
- Contact has not been possible because of the vast size of the galaxy and universe.
- We have not been searching long enough.
- Intelligent life, if it is actually intelligent, will tend to keep quiet.
- Earth is purposely isolated by some higher intelligence. This is called the *zoo hypothesis*.
- Intelligent civilizations tend to fear contact and therefore try to conceal themselves.
- Contact *has* been made and aliens are here among us.
- Contact *has* been made, but the government has suppressed the evidence.
- Aliens are so different in psychology, physiology, and technology that contact is not possible.

But these are merely speculations, and since about 1960 a few scientists have devoted themselves to resolving the Fermi Paradox and answering the question, are we alone? Astronomer Frank Drake from Cornell University performed the first *Search for Extraterrestrial Intelligence* (SETI) experiment in 1960 when he pointed a 26-meter radio telescope skyward and listened for signs of life. Soviet scientists in the early sixties also took a strong interest in SETI, and we have been listening

* *For more, see the book If the Universe Is Teeming with Aliens ... WHERE IS EVERYBODY?: Fifty Solutions to the Fermi Paradox and the Problem of Extraterrestrial Life by Stephen Webb (2002).*

continuously for the last fifty years. But apparently no one has yet heard or seen anything suggesting that life of any kind is out there. Indeed, where is everybody?

More recently another approach for finding extraterrestrial life has become promising: searching for planets outside our own solar system, and particularly Earth-like planets. Finding an extrasolar planet, or *exoplanet*, is very difficult. The light from planets orbiting even the closest stars is so faint that even with our biggest telescopes it would be difficult to detect and, moreover, it is overwhelmed by the light from the parent star. But in 1988 a team of Canadian astronomers, Campbell, Walker, and Yang pioneered the *radial velocity technique* that measures the miniscule wobble of a star caused by the tug of gravity from an orbiting planet. With this new technique they discovered the first exoplanet, orbiting the star Gamma Cephei. This finding was not confirmed until 2002, but other astronomers began to use and refine the difficult technique; by 1995 the first undisputed exoplanets were announced by Swiss astronomers Mayer and Queloz, ushering in the modern era of planet hunting.

For the next ten years or so Geoffrey Marcy and his team at the University of California Berkeley became the most prolific planet hunters, discovering seventy of the first one hundred exoplanets, and by 2011 the official count stood at nearly 600. Most of these were large Jupiter-like planets capable of tugging noticeably on the parent star, not tiny Earth-like planets where life might be found. Was there any hope of finding an Earth?

Using another technique, the answer is yes. With a sensitive enough telescope, astronomers can detect a very slight dip in the brightness of a star when an orbiting planet crosses in front of it (this is called *transiting*). But ground-based telescopes are limited in their sensitivity by the murky atmosphere of Earth, so in March 2009 NASA launched the Kepler spacecraft carrying a very sensitive telescope (called a photometer) that looks at a section of sky with about 156,000 stars in it. In 2010 Kepler began collecting data and by the end of its first year of operation 1,235 new candidate planets were reported, nearly tripling the total number of known exoplanets (Cowan 2011). Of these, 68 planets were roughly the size of Earth (1.5 to 2.0 times as big), 288 were a few times bigger than Earth, 662 were about the size of Neptune, and 165 were the size of Jupiter (up to twice as big).

Many of the 68 planets that were rocky and Earth-like orbited their star either too close or too far away for life to be possible. For a planet to support life it must orbit its star in the "Goldilocks zone" where it's not too hot and not too cold, but *just right*. More precisely, the Goldilocks zone (or habitable zone) is where liquid water can exist. In our solar system, Venus, Earth, and Mars are all in this favorable location, but Venus has surface temperatures of 800° C because of a runaway greenhouse effect, and Mars is too small to hold much of an atmosphere or water. Of the 1,235 new planets discovered by Kepler in its first year, 54 are considered candidates for liquid water and life. Five of these are small rocky planets and the other 49 are larger planets (up to the size of Jupiter) that could have rocky moons orbiting them where life could reside.

In its search for planets, the Kepler spacecraft has also detected a number of planetary systems with as many as seven planets orbiting the star. One of the most interesting (so far) is the Gliese 581 star system (Figure 3-1). Gliese 581 is a star one-third the size of the Sun, and only about 1% as bright, located 20.3 light-years away from us. There are six confirmed planets orbiting this star and one of these, named Gliese 581d, is now considered an excellent candidate for a planet that could have life – it is small and rocky, about 1.8 times the size of Earth, and orbits within the habitable zone of its star. Even though the energy input from the star is vey low, mathematical models suggest that a CO_2 atmosphere could trap enough heat to make this planet livable (Wordsworth et al. 2011).

Other planetary systems have been discovered that seem very strange, and they challenge the prevailing theories of planetary formation. The star system called Kepler-11 is the most compact yet seen. Its six orbiting planets are similar to Uranus and Neptune in composition, but orbit within a distance closer than Venus is to the Sun, and five of them orbit closer than Mercury is to the Sun. There is at present no theory of planet formation that accounts for such a configuration.

At this writing the findings from the Kepler mission are just beginning to pour in. The mission has at least another two years of data collection (it may be extended another two years) and scientists will be sifting through that data for years to come. Scientists are extremely excited and optimistic about the potential for the Kepler mission to find another Earth and to determine how common, or rare, Earths are. At this moment we are closer than ever to knowing whether life exists anywhere else in our neighborhood, but we still do not know.

Figure 3-1

Comparison of Gliese 581 System and the Solar System

Wikimedia Commons/European Southern Observatory

Early Earth

The surface of the early Earth needed to cool and then acquire several more things before life was possible: an atmosphere, oceans, and eventually a magnetic field. These three things were very interrelated, and all were tied to Earth's hot, molten interior. One of the many fortunate circumstances in the formation of the Sun and

Solar System is that Earth received heavy elements like uranium and thorium that are radioactive. Through the process of radioactive decay, elements like these produce abundant heat; this is still the heat source for the volcanoes, geysers, and hot springs we see today. The molten interior of the Earth produced volcanic eruptions that included a large amount of "out-gassing." Gases like carbon dioxide, water vapor, nitrogen, and sulphur dioxide spewed out of vents to form an atmosphere that was held tightly by Earth's gravity to form a protective envelope. This thin bubble has shielded Earth from falling debris for billions of years ("falling stars" are tiny meteors burning up in the atmosphere) and it also separates the surface from the harsh vacuum of space. The early atmosphere was dominated by carbon dioxide and water vapor, and it remained like this for almost 2 billion years before oxygen finally became abundant.

This early Earth was much hotter than today, because of the prolific radioactive decay in the interior, the greenhouse effect caused by the CO_2 atmosphere, and a constant bombardment by meteors. Most of the steam that out-gassed remained as water vapor in the atmosphere, but as Earth cooled, much of this water vapor condensed into liquid water on the surface, making the first oceans. Most scientists believe that this cannot account for all the water that fills today's oceans. So where did the rest come from?

Until recently, the most widely accepted theory was that most of Earth's water came from a place where H_2O is abundant—the outer Solar System—and it was delivered to Earth by comets. The Kuiper Belt and the Oort Cloud, far beyond the orbits of the planets, contain billions of "dirty ice balls" left over from the formation of the Solar System. Some of these fall inward toward the Sun, usually moving into a giant elliptical orbit. As they get close to the Sun they acquire long tails of material vaporized by the heat from the Sun and are seen as comets. The early days of Earth and the Solar System were characterized by intense bombardment of the planets by asteroids (small rocky bodies) and comets (small icy bodies). This *era of intense bombardment* lasted on Earth from about 4.6 to 4.0 billion years ago, and it seems plausible that during this time comets delivered the water that filled the oceans.

But when researchers analyzed the tail of the large comet Hale-Bopp in 1999, they found that the water it contained did not chemically match Earth's water (NASA Press 1999). A small amount of the water on Earth is "heavy water", denoted as HDO rather than H_2O. In heavy water, one of the two hydrogen atoms is a deuterium atom that contains a neutron in the nucleus. The water in comet Hale-Bopp had a much higher percentage of heavy water than Earth's water, so the research team concluded that comets were probably not the source of Earth's water. Then in 2001, the analysis of the break-up of the comet LINEAR showed that its water did match Earth's water, and now scientists have discovered that there are some comets that formed closer to the Sun, near the orbit of Jupiter, whose water matches Earth's water, while the comets that formed in the outer reaches of the Solar System have water that does not match (NASA Science News 2001). Planetary scientists are now rethinking the theory of comets—what they are, how they form, and what they are made of—and the study of comets remains a very important area of research. Meanwhile other theorists have proposed that Earth may have formed with all its water from the beginning, if water molecules were held tightly inside the mineral olivine (Cowan 2011). At this writing the question of how the Earth got its water is still unanswered.

The large amount of water in the oceans today would be enough to completely cover Earth to a depth of over a mile if the surface was perfectly smooth. For the first billion years after the oceans formed, they did in fact cover the entire Earth. But higher-elevation landmasses eventually formed and rose out of the oceans. Two separate processes produced these landmasses: *volcanism* and *plate tectonics*. The Hawaiian and Galapagos Islands are examples of landmasses that were formed by volcanism, and these kinds of large, circular features can be seen elsewhere in the Solar System. Huge volcanic mountains are evident on Venus, Mars, and some of the moons of Jupiter and Saturn. But long chains of mountains—like the Rocky Mountains, the Alps, and the Himalayas—have been seen only on Earth. Chains of mountains are signs of plate tectonics, a process that is apparently unique to Earth.

Plate tectonics began to evolve during the era of intense bombardment about 4 billion years ago when Earth's surface was kept molten by the many meteor impacts. The less dense molten material floated to the surface and eventually cooled and solidified into large plates. The underlying denser material remained molten because of the heat from radioactive decay, so that the solid plates above could float and move around. As these plates bumped up against each other, some plates got pushed upward, forming mountains. The landmasses we call continents are plates that thickened enough to stick out from the oceans, and they have moved around significantly over the course of Earth's history (we'll return to this in the next chapter).

A further protective feature that made Earth favorable for life was its magnetic field. This shields the surface from the intense *solar wind*—protons and electrons that stream out from the Sun and slam into Earth at speeds up to a million miles per hour. The solar wind would be deadly for life near the surface of Earth, ripping apart large molecules, if it was not stopped very effectively by our magnetic field. The mechanisms that create our magnetic field are not yet fully understood, but it is certainly dependent on the rotating inner layers that are both liquid and solid. Once again, it is radioactive decay inside Earth that produces the heat necessary to maintain liquid layers. In the earliest days of Earth, the magnetic field did not exist because interior layers had not yet differentiated. The best evidence today suggests that the magnetic field emerged by about 3.45 billion years ago (Matson 2010).

But by about 4.0 billion years ago, some 9.7 billion years after the Big Bang, the intense bombardment of Earth was subsiding, the atmosphere was thick with carbon dioxide and water vapor, and the oceans covered Earth. Out of the cosmos had emerged a planet that was ready for life. Somehow, somewhere deep in the primordial oceans, the molecules of life began to assemble.

Part One Summary

The universe was born at the moment of the Big Bang, now confirmed to be 13.75 +/- 0.11 billion years ago (NASA 2012). The first minute or so in the life of the universe is truly mind-boggling, with nearly infinite temperatures and densities. Yet cosmologists have mapped out a probable sequence of events in which the fundamental forces and particles emerged, the structural framework of the universe evolved, and the first elements were synthesized. Today the evidence continues to accumulate in support of this hot Big Bang scenario.

As the universe cooled in the first few minutes, the first elements formed—hydrogen, helium, and a little lithium—but it would be another 400 million years before things got cool enough for gravity to pull together great clouds of hydrogen into the first stars. This was a crucial development, because it is stars that create the elements heavier than helium. The beginning of star formation was one of the most significant events in the entire history of the universe: before there were stars, the universe was a primordial, chaotic, world of darkness. With the advent of stars, the universe lit up and began to take on the elegant structure we see today.

There are many types of stars, but the single thing that determines almost everything about a star is its mass. Medium-sized stars like the Sun live for a very long time, making them reliable, long-lasting energy sources for planetary systems. Large-mass stars, often called blue stars, may have masses of fifty or more times the mass of the Sun. They pour out vast amounts of energy each second, but have very short lives, so they cannot support planets with life—they simply are not around long enough. But large-mass stars do something that is very important: they create the heavy elements and cast them out into space as they die in massive supernova explosions. The remnants of supernovae are rich in heavy elements and the new stars that condense out of this material also contain these heavy elements.

When the universe was about 9 billion years old (about 5 billion years ago) a giant cocoon with thousands of stars began to form in the Orion Arm of our Milky Way Galaxy. The first star to form in this cluster was a large blue star, and then several thousand smaller stars formed, all in a space that was 5 to10 light-years across. One of these smaller stars was our Sun. Sometime just after the Sun and its planetary system had formed, the large star in our cluster reached the end of its short life and died violently in a supernova explosion. The nearby Solar System was heavily injected with all of the 92 elements up to Uranium. This gave Earth the crucial ingredients that would later make life possible.

After the large star in our birth cluster went supernova, the gravity in the cluster was diminished and the Sun and its several thousand siblings drifted apart, dispersing throughout the galaxy. The Sun found its way into the nice quiet neighborhood where we now live, about 30,000 light-years out from the center of the galaxy. This medium-sized single star formed with a system of planets revolving around it in nearly circular orbits. One of these, the third one out, was a rocky planet that circled the Sun at just the right distance for liquid water to exist; and it also had enough gravity to hold an atmosphere, unlike some of the other rocky planets.

In these early days of the Solar System, the planets were bombarded relentlessly by debris left over from the formation of the Solar System, pulled in by the gravity of each planet. The intense bombardment of Earth kept its surface molten, which allowed lighter material to rise to the surface. Some of the bombardment included comets containing water from the outer Solar System, and eventually enough water accumulated to fill the oceans.

By about 4 billion years ago, the era of intense bombardment drew to a close. As Earth cooled, the less-dense surface material solidified into giant floating plates, and liquid water condensed out of the atmosphere to cover the entire planet. Somewhere in those oceans, perhaps deep down near hot water vents, organic molecules somehow assembled into something completely new: *life*.

Part Two

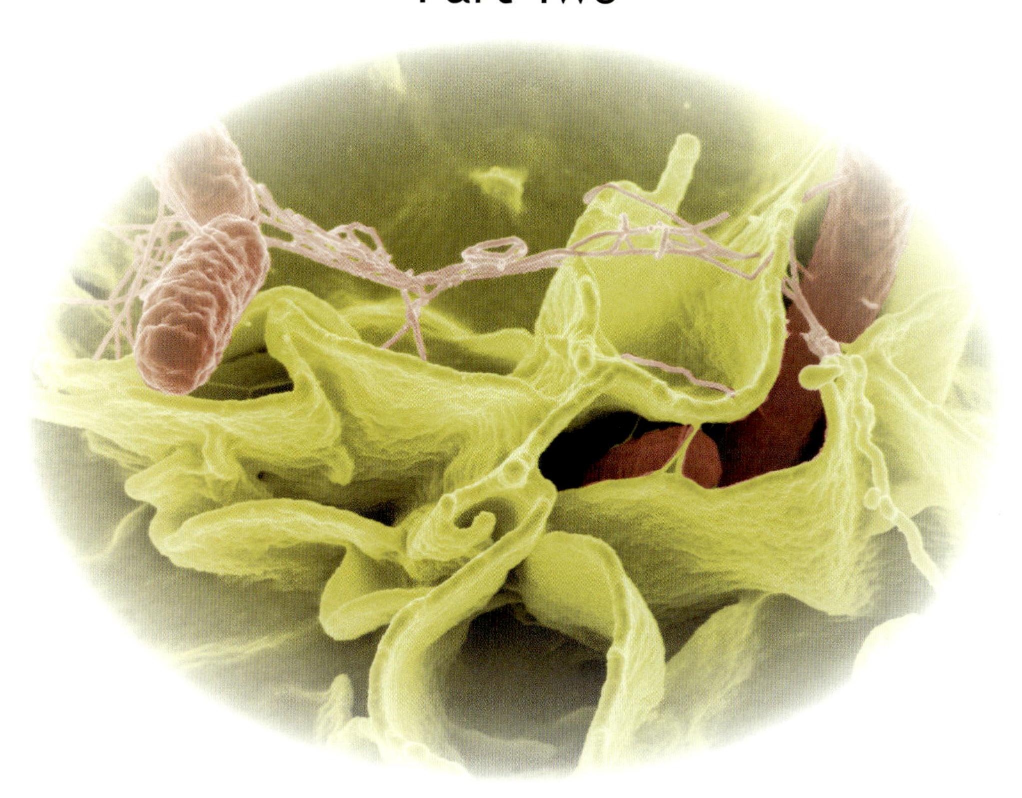

The Age of Bacteria

PART TWO

THE AGE OF BACTERIA
Life on Earth Begins

It took the universe nine billion years to make the Sun and Earth, and in its earliest days our planet was a poisonous molten inferno, completely inhospitable for life. Even so, it was the only planet in our Solar System that orbited the Sun in the "habitable zone", and it was rich in the heavy elements that would eventually make life possible. After some 600 million years, Earth cooled, the oceans formed, and conditions became favorable for life to begin. In this part of our story we will explore what life is, how it might have gotten started on Earth, and what its first 1.5 billion years was like, when bacteria ruled.

Time Context for Part Two

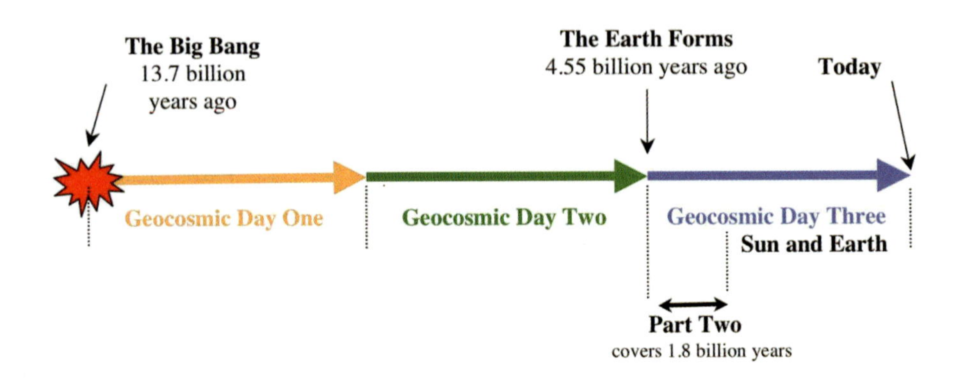

Main Events of Part Two: The Age of Bacteria

Event	Years Ago	Geocosmic Time
Earth forms	4.55 billion	Day Three, 12:00 am
Hadean Era ends	4.0 billion	Day Three, 3:45 am
First oceans	3.9 billion	Day Three, 4:15 am
Life begins (DNA is present)	3.8 billion	Day Three, 4:45 am
Earliest known fossil bacteria	3.5 billion	Day Three, 5:30 am
Earliest known photosynthesis	3.5 billion	Day Three, 5:30 am
First complex cells (eukaryotes)	2.5 billion	Day Three, 11:00 am

Chapter 4
The Origin of Life

Out of the Furnace: The Hadean Era

Geocosmic Time: Day Three, 12:00 am to 3:30 am

By about 4.55 billion years ago, the Earth had formed out of the same rotating cloud of material that spawned the Sun and the rest of the Solar System. However it would be another 700 million years or so before life was possible. These earliest days of the Earth are called the *Hadean Era* because Earth was a molten inferno, like the underworld of Hades portrayed in Greek mythology. Earth's surface was kept liquid hot by the constant bombardment of asteroids and comets, leftovers from the formation of the Solar System being pulled in by gravity. Most of the debris strewn throughout the early Solar System was swept up by the vast gravities of the Sun and Jupiter, but Earth also took a beating.

During the Hadean Era, also called the era of intense bombardment, asteroids as large as 100 kilometers routinely slammed into the Earth, generating immense heat that kept the surface molten. To make things even more Hadean, volcanic activity was widespread, spewing molten rock over Earth's surface and expelling carbon dioxide, hydrogen sulfide, steam, and methane into the atmosphere. Earth's early atmosphere had no free oxygen and would have been quite poisonous for most of today's life. Life as we know it was not possible on Earth at this time because it was far too hot for liquid water or the complex molecules of life to exist.

The bombardment also had benefits. Vital raw materials were delivered from the outer Solar System, including H_2O, amino acids, and other hydrocarbon molecules. Some scientists think that most of the water now on Earth arrived during the Hadean Era in "dirty snowballs." Some of these linger today in the outer reaches of the Solar System and occasionally fall in toward the Sun as comets. Many of the organic building blocks of life may have arrived on asteroids and comets.

By about 4.0 billion years ago, or about 3:45 am in the 3rd geocosmic day, most of the lingering debris in the Solar System had been vacuumed up by the gravities of the Sun, Jupiter, and the other planets, and the intense bombardment of Earth finally ended. Since then, for the last 4 billion years, asteroid impacts have been very rare, but not unknown. The most recent major impact event was 65 million years ago when a 10-kilometer asteroid struck the Earth near the Yucatan Peninsula. This is now called the *K-T extinction event,* and it eliminated not only the dinosaurs but also many other forms of plant and animal life. Yet, as we will see, cataclysmic events such as these are also very fortunate—the K-T event allowed mammals to fully emerge from the shadows of the dinosaur and later evolve into *Homo sapiens*.

As the period of intense bombardment came to a close, the Earth began to cool. Eventually it cooled enough for the surface to solidify and for most of the steam in the atmosphere to condense into liquid water. It may have rained torrentially for millions of years, filling the first oceans. The presence of large quantities of liquid H_2O on Earth was just one of the many critical factors that made the chemistry of life possible. One of the great unanswered questions in science today is whether there are other Earth-like planets in our galaxy where life is possible. Most scientists believe this is likely, but none has yet been found among the first thousand or so exoplanets that have been discovered since the 1990s. So far, our planet-hunting technology has limited us to finding mostly large (Jupiter-sized) planets, but better instruments, like the Kepler spacecraft, are beginning to detect smaller Earth-like planets for the first time.

Ancient sediments, preserved as exposed rock, show that Earth's oceans had formed by about 3.9 billion years ago and covered the entire planet; the first landmasses would not rise out of the oceans for another billion years. Then, not long after the oceans formed—perhaps as little as 50 million years later—something incredible happened deep in the oceans, safe from the Sun's searing radiation, probably near hot water vents where energy and raw materials were abundant: *life appeared*.

What is Life?

The earliest known life on Earth is evidenced in ancient rocks from Isua, Greenland, discovered in the late 1960s by geologists exploring for ore deposits. These are among the oldest rocks on the surface of the Earth and have been dated to about 3.85 billion years old (Tenenbaum 2002). The signs of life these rocks contain are not fossils of actual microbes, but chemical signatures in the rock: molecular fossils. When these rocks are analyzed, they are found to contain higher than normal levels of carbon-12 compared to carbon-13. When living things take in carbon (as CO_2) they have a preference for carbon-12 and therefore all living things have higher levels of carbon-12 compared to non-living things. The higher levels of carbon-12 in the Isua rocks can only be explained by the presence of primitive life. If life left these carbon fingerprints 3.85 billion years ago, it must have originated some time just before this, probably about 3.9 billion years ago.

Here, then, is one of the greatest mysteries. Before about 4.0 billion years ago, there was no possibility of life on Earth because conditions were so hostile (liquid water could not exist), but then by about 3.85 billion years ago life was clearly established. What happened during this relatively short window of time? How did life get started? Before exploring the origin-of-life question, we must understand more clearly what makes something living.

Most people would agree that a butterfly is living and a rock is not. However there are many non-living things that exhibit life-like qualities. For example, crystals grow in a glass of sugar-water as the water evaporates; a forest fire moves large distances, eating up trees and producing energy; and man-made viruses can enter a computer, move around in the hard drive, and carry out destructive tasks. Yet none of these are a form of life.

For thousands of years, people have pondered the question of what defines life and no one has yet come up with a clear, simple definition with which everyone agrees. Scientists today generally subscribe to a theory of living systems that includes the following criteria:

- ✓ Living things consume resources and energy from the surrounding environment and produce wastes, and this results in growth.
- ✓ Living things respond to stimuli. For example, single-celled organisms recoil from something hot; a deer runs away from the sound of a gunshot.
- ✓ Living things maintain their internal environment at very specific conditions. For example, humans maintain a body temperature of 98.6°F. Biologists call this *homeostasis*. A membrane of some kind separates the inner and outer environments.
- ✓ Living things reproduce. They make offspring that are similar to themselves.
- ✓ And, perhaps most definitively, all forms of life that we know of are based on the self-replicating, information-storing molecules, DNA (deoxyribonucleic acid) and its companion RNA (ribonucleic acid).

Science and Discovery

DNA and RNA

Chromosomes were first seen in the 1840s by biologists studying plant cells with a microscope. Most cells they looked at had a nucleus containing chromosomes, but they had no idea what these structures did, or what they were made of. It took nearly 100 years to establish that chromosomes are made of DNA and function as the master control and information storage system for every living organism.

It was a Swiss chemist, Frederick Miescher, who first identified DNA in 1869, but he thought it was little more than "packing material" in the cell nucleus. Since then biologists have gradually put together an astonishing picture of what this molecule does, from its central role in the origin and evolution of life to its function as the machinery of life itself. Today DNA research is one of the most active and productive areas of science. Scientists are learning more about DNA all the time, yet it appears that we may be just scratching the surface.

Inside each human cell is a cell nucleus containing 23 pairs of chromosomes (Figures 4-1 a and b), and these chromosomes are made up of long strands of coiled up DNA – a total length of about *six feet* spread over the 23 chromosomes! We now understand that the DNA molecule acts as a very long string of information. Our 23 chromosomes could be thought of as an encyclopedia with 23 volumes – all together we call it the human genome. If we could take all of the information in one volume of an encyclopedia and string it into one very long sentence, this would be like a long strand of DNA coiled up in one chromosome. In fact there is much more information stored in the DNA of one chromosome than in one volume of any encyclopedia.

Figure 4-1a
The genetic machinery, from chromosomes to DNA

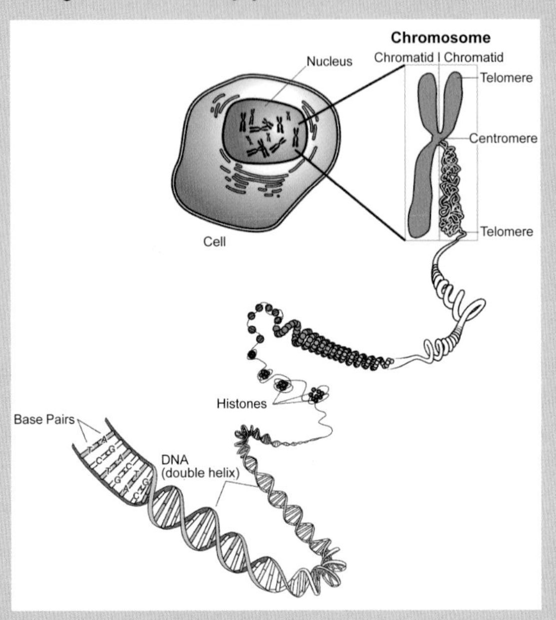

Courtesy: National Human Genome Research Institute

Figure 4-1b
The human genome, with 23 pairs of chromosomes. This is a male, with XY for the 23rd pair of chromosomes. In a human female the 23rd pair is XX.

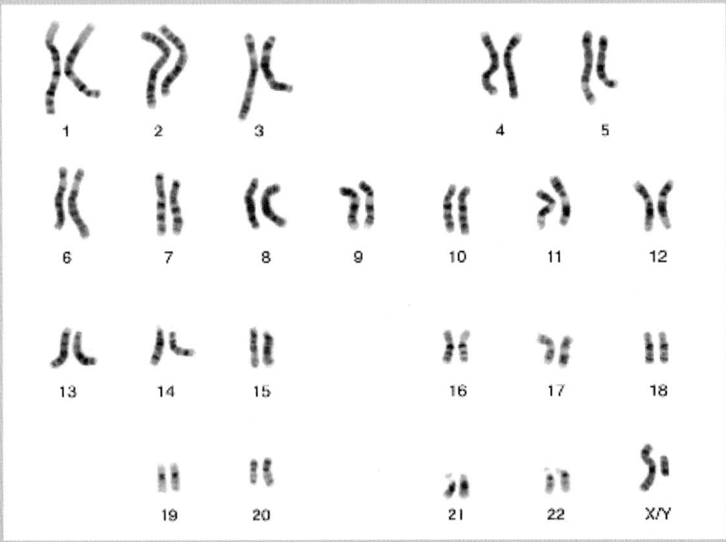

1 2 3 4 5
6 7 8 9 10 11 12
13 14 15 16 17 18
19 20 21 22 X/Y

Courtesy: National Human Genome Research Institute

Encoded in our DNA is a detailed set of blueprints for every aspect of our body, from individual molecules to eye color, as well as complete operating instructions for everything our body does. Our DNA is something like a computer hard drive storing vast quantities of information *and* a sophisticated operating system that manages all that information. Researchers are still discovering deeper layers of information management in the DNA molecule, and it's now clear that our most advanced super-computers are quite crude by comparison; yet this molecule is so tiny we cannot see it with a microscope. How is all this possible?

Figure 4-2
A schematic representation of the DNA molecule

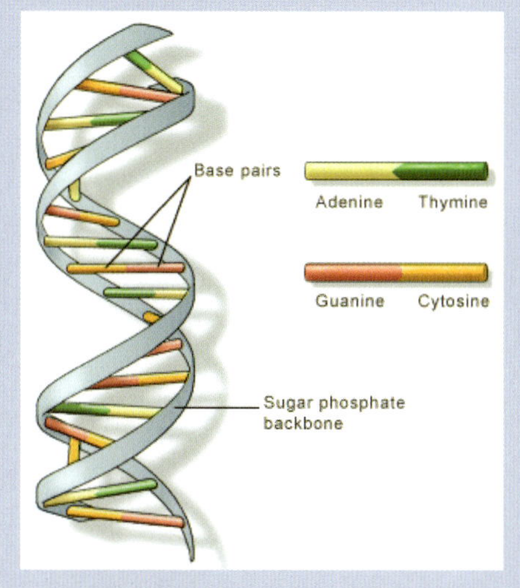

U.S. National Library of Medicine

The first big steps toward understanding how DNA operates came in 1953 when Francis Crick, James Watson, and Maurice Wilkins unraveled the chemical structure of DNA using the x-ray images of the molecule that were taken by Rosalind Franklin. They discovered that the DNA molecule is like a very long ladder, twisted along its length in a shape they called a "double-helix" (Figure 4-2). The sides of the ladder are a repetitious string of sugar phosphate molecules, and it's the rungs of the ladder that store information.

Each rung is made of two molecules that join at the middle of the rung. These molecules are called *nucleotides*, and are also referred to as *bases*. So the rung of the ladder is sometimes called a *base pair*. There are four kinds of nucleotide bases that can pair up to make a rung of the DNA ladder, and these are denoted as **A**, **T**, **C**, and **G** (short for Adenine, Thymine, Cytosine, and Guanine). However these four bases can only pair up in certain ways: **A** can only pair with **T**, and **G** can only pair with **C**. So there are only four types of rungs, and they are:

A-T
T-A
G-C
C-G

We can think of a DNA molecule as the long sequence of letters strung out along one side of the ladder (the other side is duplicate information, acting as a protective back-up system). This is an alphabet with four letters: **A**, **T**, **C**, and **G**; and this alphabet is used to make three-letter words, called *codons*. Some possible codons would be:

ACG, **TAC**, **CCT**, **GCT**, **ACA**, and so on.

If you wrote out all combinations of 3 letters from the alphabet of four letters, you would find that there are 64 possible codons – so there are 64 possible words in the DNA language. Life on Earth uses only about twenty of these words, but even twenty words can be strung together in an infinite number of ways—in very short sentences or very long, and in any order.

What do these words and sentences mean? James Crick offered an answer to this in 1958, proposing what came to be called the "central dogma" of molecular genetics (he later regretted the use of the word *dogma* rather than *theory*). This theory says that DNA transfers blocks of its information to a sister molecule called *messenger RNA* (or mRNA, see Figure 4-3). The mRNA then moves out of the nucleus, carrying this information to another part of the cell where structures called *ribosomes* use the information to build proteins. In shorthand, Crick's idea can be symbolized as:

DNA ----> RNA ----> Protein

The key to this process is in the meaning of the words and sentences contained in DNA. Each 3-letter word (or codon), made from three rungs on the ladder, represents an amino acid. More accurately, the shape of the molecules that make up the codon fits with the shape of a particular amino acid. So amino acids get attached to a particular codon, and then get assembled in the order of the codons on a section of DNA. A string of codons specifies a string of amino acids; and a string of amino acids is a protein. So we say that a certain stretch of DNA *codes* for a particular protein. And proteins do just about everything in living things. A DNA sentence, which is often called a *gene*, contains the instructions for building a specific protein by stringing together a series of amino acids. In more complex life forms like humans, a single gene can build many proteins that function together, and some parts of a gene contain instructions that turn the gene on or off.

Generally a gene can be thought of as a discrete piece of DNA that controls one particular aspect of the organism. For example, a stretch of DNA has recently been identified on human chromosome number 19 that makes a protein that helps the body remove excess cholesterol. When there is an error in this gene—when just one letter is incorrect—the protein is not made properly and the person is then more vulnerable to the clogging of the arteries caused by excess cholesterol. However, for many diseases there are multiple genes involved, located on different chromosomes, which interplay in complicated ways. Biologists still debate exactly what is the best definition of a gene.

Proteins are huge, complex molecules sometimes made of hundreds of amino acids. They fold and unfold intricately as they do their job and they are involved with nearly every function of living things. Some proteins are the structural components of cells, while others are the enzymes that stimulate crucial chemical reactions, or the hormones that communicate within an organism, or the antibodies that latch onto foreign invaders like viruses and bacteria. The simplest life forms (like bacteria) build and use about 1000 different proteins, while human DNA codes for more than *sixty thousand* different proteins that are all needed somewhere in our body.

The process of building proteins begins when one section in a long strand of DNA unzips—that is, the rungs of the ladder come apart in the middle where the two bases join. Then a complementary strand of RNA is built by filling in the other half of the ladder. Every **C** on the DNA-side of the ladder will find a new **G**, every **A** will find a **T**, and so on, until a new strand of RNA is finished. We call this new strand messenger RNA or mRNA. RNA looks like just one side of a twisted ladder (see Figure 4-3), but it still contains all the information in the original DNA strand. When the strand of RNA is complete—when it has captured a full sentence from the DNA—it detaches and moves away while the DNA molecule zips back together. This process of transferring information from DNA to mRNA is called *transcription*.

Figure 4-3
A comparison of DNA and RNA

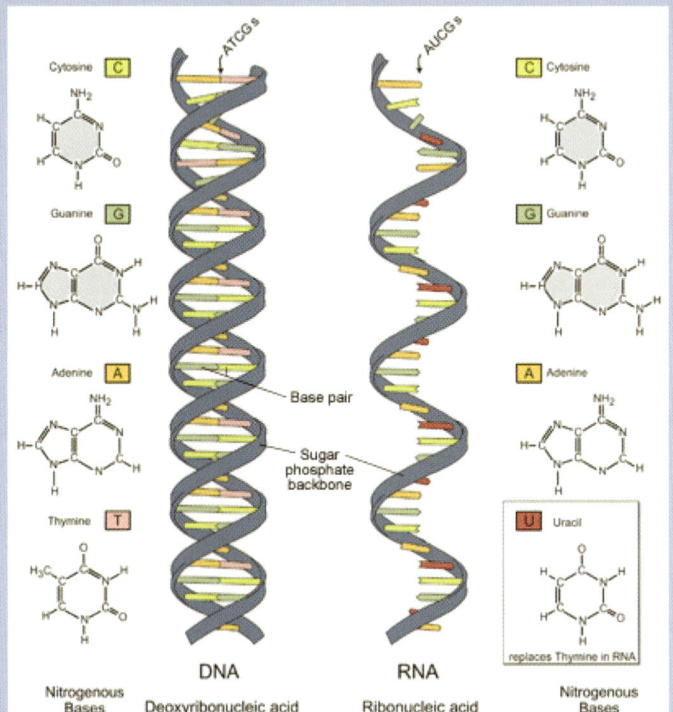

Courtesy: National Human Genome Research Institute

The newly made RNA strand then moves out of the cell nucleus to a ribosome, where the sentence is used to actually string together the proper amino acids to build a specific protein. Ribosomes are protein factories. This part of the process, where the information carried by mRNA is turned into a protein is called *translation*. The journey from DNA to protein, using transcription and translation, is called *gene expression*. When a protein is completed it can then move away to fulfill its mission somewhere in the cell or even outside the cell in another part of the body. The DNA-RNA complex is the machinery of life, churning out proteins for nearly everything a living thing does.

The double-helix structure of DNA also makes self-replication possible. DNA can make nearly perfect copies of itself by unzipping from one end. Each half of the ladder then builds a complementary strand, with an **A** always finding a **T**, a **C** finding a **G**, and so on. In this way, the original DNA strand that unzips becomes two new strands almost exactly like the original. There are occasional copying errors and some of these are corrected, but some remain. The uncorrected errors, called *mutations*, play an important role in the evolution of life.

The entire genome of a plant or animal – the entire length of DNA spread out over the chromosomes—self-replicates every time a new cell is made (generally called mitosis). After the original DNA has self-replicated, the two sets of DNA separate and move apart. Eventually a membrane forms as the original cell divides into two, and the two duplicate sets of DNA end up in the two new cells. This is happening all the time in our body as new tissue grows or is repaired. In a matter of minutes a cell can duplicate the vast library of information contained in our DNA and insert the copy into a new cell. DNA relentlessly perpetuates itself by replicating over and over, until the organism that houses it dies. Yet the information in our DNA transcends even death, because copies of the information are passed along to offspring. Life goes on as long as DNA survives and is passed on to new generations.

Our discussion of DNA so far represents the knowledge level of biologists around 1960. Since then, DNA research has flourished, producing more new knowledge than any other area of science. We are now able to read the information coded in DNA and determine what specific sections of DNA actually do; we can alter sections of DNA, and insert pieces of DNA from one organism into another. This brings the promise of eliminating devastating diseases, but it also raises difficult ethical questions. Are we playing God? Will there be unintended consequences?

Until recently, the prevailing view of DNA was that it was segmented neatly into protein-coding genes, in the simplest case, one gene codes for one protein that has one specific function in the organism. The DNA in a chromosome was thought to be a long string of genes neatly ordered like the cars in a freight train. As molecular biologists began to decipher the human genome in the 1990s, they discovered that only about 2% of the genome consisted of protein-coding genes and the other 98% seemed to do nothing, so it was dubbed "junk DNA." However it has been discovered more recently that the other 98% is doing something very important: rather than transcribing into proteins, it is transcribing into a wide variety of RNAs that have important regulatory roles. There are now more than 20 classes of RNA known, and one of the most important of these is the micro-RNAs, or miRNAs, first discovered in 1993 (Lee). These are short strands of RNA with only about 20 or so nucleotides, but they are involved with nearly every piece of cellular machinery. They regulate gene expression, not only turning genes on or off, but also acting like a dimmer switch that fine-tunes the degree of expression (Mourelatos 2008). They also monitor things like temperature, chemical conditions, electrical balance, and other signals from the environment, and then tell the cell how to respond. Researchers have now found that human brain function is facilitated by specific miRNAs manufactured in neurons (brain cells) (Medical News Today 2006). Micro-RNAs could help explain how the human brain can process so much information so quickly.

Twenty years ago it was believed that the main function of DNA was to turn its stored information into proteins, using RNA as a bridge to the protein factories (the ribosomes). Now it seems that protein-coding is only about 2% of the job, and making micro-RNAs is a much bigger part of what DNA does. The discovery of micro-RNAs has brought a revolution in our understanding of the machinery of life. We now know that DNA contains even more information than previously thought and this information is layered and intermingled in nonlinear ways. RNA has become the new hero of cellular function.

The DNA/RNA complex is at once the *library* of information, the *factory* that turns the information into something real (proteins), and the *overseer* that monitors and regulates the processes (using proteins and RNAs). Included in the library of information are complete blueprints and operating instructions for every aspect of an organism and the means to carry out those instructions. It has been estimated that a Boeing 747 jetliner, with 6 million parts, has about the same level of complexity as a yeast cell (Vines 2002). Yet the entire set of blueprints and operating instructions for something as complex as a 747 is contained in the DNA of a lowly yeast cell. The human body is much larger and more complex, but still the complete set of plans and instructions is contained in our 23 chromosomes that reside in every one of our cells. With this picture in mind, let's return to our story at the point where life on Earth began.

The Origin of Life

Geocosmic Time: Day Three, about 3:30 am

It was 3.9 billion years ago and oceans covered the entire Earth to a depth of half a mile. There was no dry land because the continents had not yet risen, but the oceans were rich in the raw materials of organic chemistry, ranging from simple dissolved salts and minerals to larger hydrocarbon molecules like sugars, lipids, nucleotides, and amino acids. Hot water vents on the ocean floor and massive electrical discharges on the surface could have supplied energy to drive the chemistry toward larger and more complex molecules; and eventually, somehow, out of this primordial organic stew, the DNA molecule assembled and life as we know it began. It is almost certain that this happened deep underwater, because near the surface of the ocean the solar wind radiation would have destroyed these fragile molecules. The Earth's magnetic field, which now shields us from the deadly solar wind, had probably not yet formed when life began. How, then, did life arise from inanimate matter?

Origin-of-life researchers have broadly identified three biochemical mechanisms that had to be developed in order for life to begin:

1. An enclosing membrane or vesicle that could separate and protect the machinery of life from a hostile and changing surrounding environment.
2. A sustainable internal source of energy; that is, a metabolic system.

3. A genetic system that could store information, direct the construction of proteins, regulate all of the other systems of life, and replicate itself.

Today there are many theories that attempt to explain how these three mechanisms could have developed, thereby transforming non-living matter into living organisms that reproduce and evolve. There is, however, no single theory that is widely accepted; the origin of life is still an unsolved problem, and a very active area of scientific research.

Some theories assert that the membrane had to develop first, so that metabolism and genes would have a protected place to emerge. These theories are sometimes called *lipids first* because cell membranes are based on lipid molecules. Other theories hold that metabolic systems had to develop first so that energy was available to power the other developments. These are the *metabolism first* theories. The *genes first* theories propose that RNA and later DNA had to develop first so that proteins could be built, information could be stored and retrieved, and self-replication could begin. In this view, external energy sources were used at first and then, later, internal metabolic systems would have emerged.

Many researchers feel that all three of these had to somehow develop together, not in isolation or in one particular order. Still, no matter which of these mechanisms we consider, the same questions arise. How do prebiotic molecules such as amino acids and lipids, molecules that were certainly present on early Earth, assemble into complex structures and systems like RNA or a metabolic cycle? Was it random chance that these complex configurations could assemble? Or, was there some driving mechanism?

Perhaps the most difficult origin-of-life problem is how the genetic system—RNA and DNA—could assemble. DNA and RNA are *polymers* made of long repeating chains of smaller molecules called *monomers*. Many of the molecules of life are polymers: proteins are long chains of amino acids, simple carbohydrates polymerize to form complex sugars and starches, and nucleotides link into long chains of RNA and DNA.

Most origin-of-life researchers today believe that DNA was probably the final version of a series of molecules that had the ability to store information and self-replicate. It is now broadly theorized that RNA came first in the earliest days of life. RNA is much like DNA but with a single strand, or half-ladder, rather than double strands (see Figure 4-3). According to this theory, there was a period of time on Earth when life was based on RNA—an *RNA world*. This primitive form of life may have been a naked strand of RNA floating in the ocean, or a collection of RNAs somehow bound together, or RNA accompanied by other structures like an enclosing membrane or a mineral surface. This RNA ensemble is thought to have been capable of evolving into DNA.

Where did the RNA come from? This is an even more difficult question because there are a number of problems with creating RNA. The first is a chicken-and-egg problem—which came first, proteins or RNA? RNA directs the manufacture of proteins; it produces proteins. However it can only do this when stimulated (or catalyzed) by certain enzymes, which themselves are proteins. So which came first, the RNA or the proteins that make it function? This question was partially answered in the 1980s when Thomas Cech and Sidney Altman discovered that RNA itself could act as a catalyst in the building of proteins (they won the 1989 Nobel Prize in Chemistry for this work). Yet this did not explain how RNA originated.

Another problem with creating RNA is that it is both complex and delicate. Some of its components, like ribose and the pyrimidines, are difficult to make and they decompose easily. One theory is that RNA had a predecessor, dubbed pre-RNA. Though its actual structure is unknown, it would have been a simpler kind of self-replicating informational molecule, and this form of pre-RNA may have had even simpler predecessors. From this viewpoint, life on Earth began with the first *self-replicating informational molecules*, and these eventually evolved into DNA. The origin-of-life question becomes a question of how the first self-replicating informational molecule formed. How could such a molecule arise from the primordial soup of organic molecules that filled the oceans?

In 1953 Harold Urey and Stanley Miller sought to answer that question in a now-famous experiment at the University of Chicago. They attempted to create the environment of early Earth in their lab. There was a warm "ocean" of liquid water in a flask that was heated so that water vapor rose into an atmosphere of methane, ammonia, and hydrogen. Then they exposed the atmosphere to a continuous electrical discharge that simulated lightening. After this ran for several days, Urey and Miller analyzed the contents of the mock ocean and found that a number of organic monomers had formed, including several amino acids.

This was an exciting result and several newspapers initially reported that life had been created. That was far from the truth, but still the results of the initial Miller-Urey experiment seemed promising as a model for how life began. It appeared that with a little tweaking of the initial ingredients and conditions, this sort of scenario could produce the molecules of life. Since then many types of "Miller-Urey" experiments have been done, and some have even produced *ATP* (adenosine triphosphate), the universal energy molecule of life. However no one has yet seen monomers assembling into polymers resembling RNA.

Some researchers have theorized that polymers could have formed on exposed clay surfaces deep under the ocean, protected from the disruptive effects of the intense ultraviolet radiation and solar wind at the surface of the ocean. The charge distribution on the clay surface could have attracted and concentrated monomers and guided their linkage into chains. Yet so far this remains only a theory, with little supporting evidence.

Another possibility emerged in 1969, when scientists analyzed a meteorite that fell near Murchison, Australia. They found that it was a rare type of meteorite that was probably part of a comet because it was 12% water. More interestingly, it was found to contain a variety of lipids and 92 different amino acids, only19 of which are found on Earth. The remaining 73 amino acids have never been seen on Earth and must have originated from other parts of the Solar System. The Murchison meteorite showed that some organic polymers like amino acids existed elsewhere, and could survive extreme cold and vacuum, as well as the intense heat of entering Earth's atmosphere.

This opened the possibility that life was seeded from sources beyond the Earth. During the period of intense bombardment that preceded the appearance of life on Earth, the materials of life, or perhaps life itself, could have arrived. This has come to be called the *Panspermia Hypothesis*, but it has few supporters among mainstream scientists. Even though the Murchison meteorite showed that organic molecules like amino acids could arrive from space, it seems highly unlikely that DNA or RNA could have arrived this way because they are such fragile molecules.

Recently, astronomers have identified organic molecules in interstellar dust clouds where new stars are born—molecules like water, ammonia, formaldehyde, hydrogen cyanide, and cyanoacetylene that could have been precursors for life. It is now clear that the basic raw materials for life are found widely throughout the universe and were likely present in the oceans of the early Earth. The pieces were apparently there about 4 billion years ago, but how did they assemble into life?

Many researchers have noted the *self-organizing* nature of the universe. Stars, with abundant hydrogen and helium that was produced in the hot Big Bang, manufacture heavier elements through nuclear fusion in their cores, elements like carbon, oxygen and nitrogen. In the environments of stars, these elements also assemble into small molecules like H_2O, CO_2, CO (carbon monoxide), NH_3 (ammonia), CH_2O (formaldehyde), and CH_3 (methane). This "spontaneous assembly" simply follows from the laws of physics and the rules of chemistry (which derive from the laws of physics); this is no more mysterious than rust (iron oxide) "spontaneously assembling" when iron and water meet.

The laws of physics are merely descriptions of the universe: *what* it is like, *what* it does, *how* it works, and even *when* something will happen. Yet scientists cannot explain WHY the universe is the way it is. The answer to the "why?" is, for the time being at least, personal. The answer to "why?" must sound something like "God made it this way" or "this is just how it came out." Even if we do not know why, the laws of physics do accurately describe the behavior of the universe and they consistently predict what will happen. The laws of physics predict that simple molecular systems can and will self-organize into more complex systems, and experiments in the biochemistry lab clearly show this.

Beginning in the 1950s a number of scientists began to put together a theory of self-organizing molecular systems. This was solidified in the early 1990s by Christian de Duve in his book *Blueprint for a Cell* (1991) and Stuart Kauffman in his book *Origins of Order: Self Organization and Selection in Evolution* (1993). They proposed that self-catalyzing (*autocatalytic*) mixtures of polymers can give rise to increasing complexity. At the simplest level an autocatalytic reaction would look like:

Reaction 1: $A + B \rightarrow C$

Reaction 1a: $A + B \xrightarrow{C} \text{more } C$

This means that reactants A and B combine to form product C, and C will then feed back into that reaction to further the production of C from A and B. More of C is produced in the presence of C, so C is a catalyst for its own production. It is autocatalytic.

Both theoretical and experimental work in biochemistry confirms that autocatalytic cycles and collective autocatalysis emerge in large sets of polymers. Writing in 2006, Kauffman says

> ... imagine two polymers, A and B, where each catalyzes the formation of the other out of fragments of the other. That is collective autocatalysis. No molecule catalyzes its own formation, rather the set as a whole is collectively autocatalytic, and achieves catalytic closure. Cells are collectively autocatalytic today....in simple models, as the diversity of polymers increases, so many reactions are catalyzed that autocatalytic sets form spontaneously with high probability.

In other words, diverse mixtures of polymers, under the right conditions, can self-organize through autocatalytic processes into more complex molecules. To put it even more simply, organic chemistry has the ability to spontaneously produce increasingly complex molecules, merely by following the natural laws of physics. Kauffman calls this "order for free."

The theory of autocatalytic self-organization makes it at least plausible that a prebiotic stew on the early Earth could self-organize into a metabolic or genetic system. Scientists in the lab have now observed hollow spheres of bilayered lipids, or liposomes, grow and divide spontaneously like bubbles. From a "lipids first" view, a vesicle such as this could be the membrane that enclosed and protected an environment where autocatalytic networks could self-organize into the machinery of life.

However it's a long way from a prebiotic soup of monomers to a full-blown self-replicating informational polymer like RNA or DNA. The late Leslie Orgel, one of the world's leading origin-of-life researchers, summed up the mystery of how life began in this way:

> Whether RNA arose spontaneously or replaced some earlier genetic system, its development was probably the watershed event in the development of life. It very likely led to the synthesis of proteins, the formation of DNA, and the emergence of a cell that became life's last common ancestor. The precise events giving rise to the RNA world remain unclear. Investigators have proposed many hypotheses, but evidence in favor of each of them is fragmentary at best. The full details of how the RNA world, and life, emerged may not be revealed in the near future (Orgel 1997).

Exploring Deeper

Does Life Violate the Second Law of Thermodynamics?

In our everyday experience we usually see things getting more disordered. Your bedroom tends to get messy—it never organizes itself. A drop of ink added to a glass of water always spreads out—it never gathers up and concentrates into something smaller. A less obvious example of this increasing disorder is the fact that heat naturally flows from hotter things to colder things. A cup of hot coffee always gets colder when we set it outside on a chilly day—we would never expect to see it get hotter after sitting out in the cold.

All of these are examples of the Second Law of Thermodynamics, one of the most fundamental laws of the universe. It can be stated in many ways but one would be:

In any closed system (where nothing can enter or leave)
the amount of disorder will always tend to increase over time.

Now young Orville has a scientific reply to his mother when she asks, "Why can't you keep your bedroom more organized?" "Mom, we're fighting the Second Law of Thermodynamics!"

Another word for disorder is *entropy*, so we say that entropy is always increasing. The most dramatic example of increasing entropy is the expanding universe. The increasing entropy of the universe even gives time its arrow. We're moving forward in time when we see entropy increasing, and we can never move backward in time because entropy would have to decrease.

The relentless march toward disorder that we see everywhere in the universe does not seem consistent with the phenomenon of life. A DNA molecule assembling from the chaos of a primordial soup 4 billion years ago is an incredible leap in organization and order. And from there, life only got more complex and ordered as evolution produced plants and animals. Some people have concluded that life is a clear violation of the 2nd Law of Thermodynamics and therefore a supernatural Creator must be the force behind the mystery of life.

However, systems *can* become more ordered—if energy is added. Certainly your bedroom *can* become organized with the input of your energy straightening it out. And heat *can* flow from cold things to hot things if energy is added. A refrigerator does just this: we make heat flow "uphill" from the cold interior to the warm room outside, but the refrigerator uses energy (quite a bit!) to do this.

The catch is that a living organism, or even the whole Earth, is not a closed system. Energy does flow into it. In the earliest days of life on Earth, hot water vents on the sea floor probably supplied this energy, and later the sun's energy became the power source of life.

From the beginning, acquiring energy has been a central issue for life. The first few hundred million years in the story of life was mostly about improvements in energy systems and, ultimately, how to tap into the sun's nearly unlimited supply of energy. Living things must have an energy source so that they can keep turning chaos into order.

Chapter 5
Prokarya: The Greatest Survivors

First Life: Prokaryotes

Geocosmic Time: Day Three, about 4:00 am

After DNA and RNA had formed and began working together to manufacture proteins, a simple cell, or *protocell,* must have evolved. The development of membranes that could maintain an interior environment was an important innovation. The first enclosing membranes may have formed in much the same way oil bubbles form in water. The first protocells were probably no more than strands of DNA and RNA, floating in a liquid cytoplasm of some kind, along with some ribosomes where proteins could be manufactured, all enclosed in a simple membrane. The design improved over millions of years to produce a form of life we now call *bacteria.*

A bacterium is the simplest kind of cell, with an outer membrane but no cell nucleus inside to enclose the DNA that floats freely in a tangle of strands. The tiny first cells of life had very few other structures (perhaps ribosomes) inside. Biologists call these simple organisms *prokaryotes.* Prokaryotes reproduce by an asexual process, *binary fission,* in which one organism splits into two nearly identical organisms. True sexual

Figure 5-1
Prokaryotic Cells
Left: Color-enhanced photo of a single cyanobacteria cell, like the early prokaryotes on Earth. The filaments are photosynthetic structures and the red dots are food storage granules. There are no cell nucleii or other organelles.
Right: Salmonella bacteria invading cultured human cells

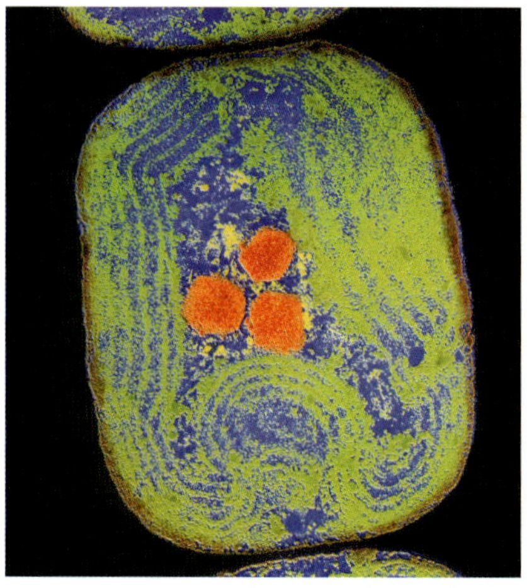

Dr. Kari Lounatmaa/Science Photo Library

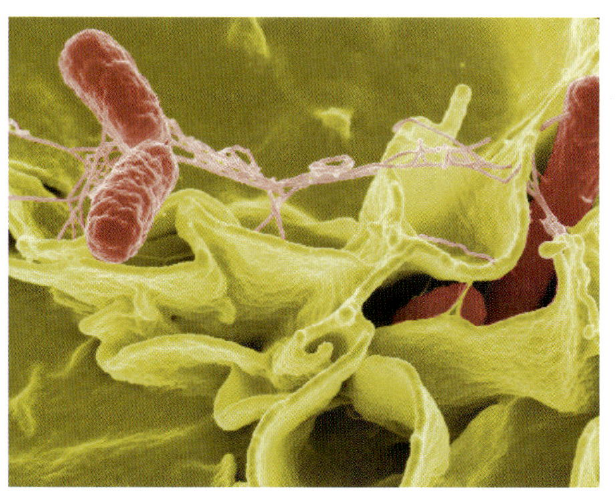

Rocky Mountain Laboratories, NIAID, NIH

reproduction, in which two different organisms combine and mix their DNA, was more than a billion years away.

The prokaryotes, which are mostly bacteria, are the most ancient form of life on Earth, and if longevity counts, the most successful. They dominate the biosphere even today, comprising some 70% of the biomass on Earth. In a single handful of dirt there are more bacteria than the total number of people who have ever lived. There can be as many as one hundred trillion bacteria of 400 different types in our digestive tract alone, and digestion would be impossible without them. They are found deep underground where there is no oxygen or sunlight, and in hostile environments of extreme temperature, acidity, or salinity.

Prokaryotic life exploded during the first billion years of life on Earth and eventually split into two separate groups, now known as *Bacteria* and *Archaea* (the *Archaea* were not recognized until the 1970s). If you could visit Earth in those days, searching for life, you would find nothing more than algae-like slime floating beneath the surface of the ocean. Many scientists now believe that this may be a fairly common scene on planets throughout our galaxy. Even in our Solar System, on moons like Titan and Enceladus, oceans of water or methane may exist that could harbor primitive forms of life. Whether this kind of primitive life could evolve into complex life, like the large plants and animals we have on Earth today, is another question. Many scientists suspect that complex life may be rare.

Most origin-of-life scientists believe that early life must have formed deep in the oceans, protected from the intense solar wind and ultraviolet radiation from the Sun. The biomolecules of life could not have formed in the presence of this radiation, which literally blows molecules apart. Today we are protected from the Sun's UV radiation by the ozone layer, an aspect of our oxygen atmosphere, and from the solar wind by Earth's magnetic field. However, the early Earth did not have an oxygen atmosphere. It was dominated by carbon dioxide, which does not provide UV shielding. It would take more than a billion years for the oxygen atmosphere to form (thanks to the prokaryotes), and several hundred million years for the magnetic field to arise.

The only likely place where the fragile molecules of life could have come together was in the depths of the oceans, safe from the brutal surface radiation, and many scientists now believe that underwater geysers, or hot water vents, were the ideal nurseries for the first life. These hot spots would have provided the chemicals, the energy, and the protection for early life. This theory was bolstered in the 1970s when new forms of primitive life were discovered in extreme environments, like the geysers of Yellowstone National Park, deep-sea hydrothermal vents, and extremely salty ponds—places where no one thought life would be possible. These "extremophyles" were originally considered to be bacteria, but DNA analysis later showed that they were a distinctly different form of prokaryotic life. Carl Woese from the University of Illinois did pioneering work in this area, and in 1977 proposed a new category of life, the *Domain Archea*. Today, most biologists agree that all of life can be divided into three great domains: Bacteria, Archea, and Eukarya (complex life, like animals and plants). The evolutionary relationship between these three domains is shown in Figure 5-2. Notice that there is a question about how the much more recent Eukaryans were related to the other two domains of life. We will return to this question in the next chapter.

By about 3.8 billion years ago life was well established but it lived deep in the oceans near hot water vents where there was no sunlight, and it was entirely anaerobic (it lived without oxygen because there was not yet any free oxygen on Earth). From the beginning, life required a source of energy and at first the source was probably a combination of heat from hot water vents and some form of simple anaerobic respiration in which a series of chemical reactions could produced a small amount of *ATP*, the molecule that all life uses to store and release energy. However these first energy systems were very inefficient and relied on extracting scarce raw materials like hydrogen, iron, phosphorus, or sulfur directly from the surrounding environment. As these "foods" were depleted, many early prokaryotes may have gone extinct. A better energy system was needed, and apparently it was developed by one kind of bacteria we now call the *cyanobacteria*, still one of the largest and most important groups of bacteria today. This better energy system used sunlight as an energy source and it eventually transformed the Earth, making complex life possible. We call this new energy system *photosynthesis*.

Figure 5-2
The Three Domains of Life on Earth

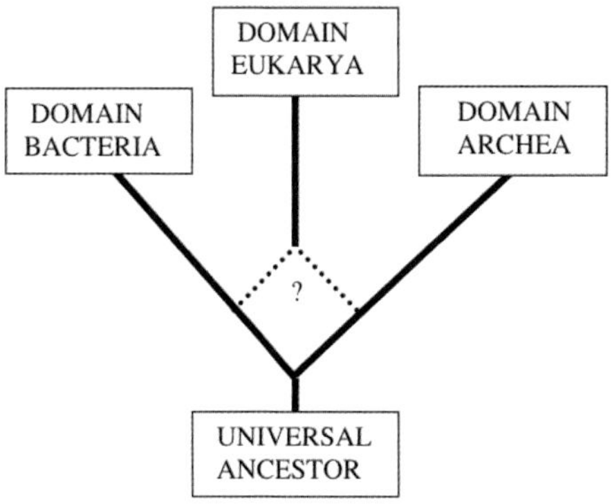

DOMAIN EUKARYA
DOMAIN BACTERIA
DOMAIN ARCHEA
?
UNIVERSAL ANCESTOR

Harnessing the Sun: Photosynthesis

Geocosmic Time: Day Three, 4:00 am to 5:30 am

Very early in the history of life, some organisms probably left the security of hot water vents in the dark depths of the ocean and moved upward toward the sunlight. They had to develop a tougher outer membrane for protection from the sun's intense ultraviolet radiation, and somehow this membrane developed the ability to capture the sun's energy, probably using pigment molecules like chlorophyll, and turn it into usable chemical energy. This was the advent of photosynthesis.

Today, the primary source of energy for nearly all life on Earth is the sun. However only photosynthesizing organisms like today's green plants can capture that energy directly. Organisms like animals can then use that energy in a secondary fashion when they eat plants. Even the energy stored in fossil fuels like coal and oil, which humans rely on so heavily today, was originally solar energy captured by plants through photosynthesis. Coal and oil are the result of plant matter being buried and compressed for millions of years, thus concentrating the energy that was captured from the sun.

Besides providing energy, photosynthesis serves another critical purpose on the Earth: it takes CO_2 out of the atmosphere and releases O_2 into the atmosphere. It was the photosynthesis of the cyanobacteria that produced Earth's oxygen atmosphere, and in more recent times (the last half-billion years) plants and animals have had a remarkable complementary relationship that maintains the balance of CO_2 and O_2: animals take in O_2 and release CO_2 while plants do essentially the reverse.

It is not an overstatement to say that complex life on Earth would not be possible without photosynthesis, so its development is one of the most crucial events in our story. Had the early life forms been limited only to chemical sources of energy, life would have had a very limited future—most likely a dead end, as chemical resources were depleted. By tapping the renewable and virtually unlimited energy of the sun, life took a path that could be sustained for billions of years.

The full-scale photosynthesis that occurs in plants today is extremely complex (Figure 5-3) and is still a very active area of research. One of the most exciting recent discoveries is that some of the energy transfer processes in photosynthesis are quantum processes that utilize wavelike behavior in systems of large molecules (Engel 2007; Collini et al. 2010). This opens the possibility that many processes in living things may be quantum in nature rather that just classical chemical reactions.

Figure 5-3

Two depictions of photosynthetic processes as they occur in plants, algae, and cyanobacteria.

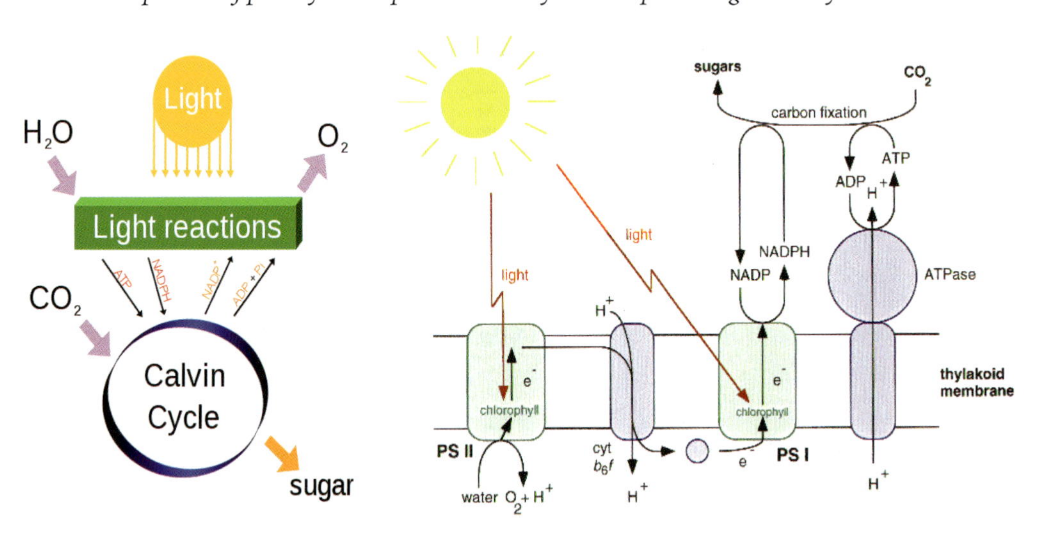

Wikimedia Commons

Courtesy: Vim Vermass, Arizona State University

Biochemists have identified two main parts to the process which they call *photosystem I* and *photosystem II* (PS I and PS II for short), but for our purposes we may simplify the process by writing:

CO_2 + H_2O + sunlight ----> sugars + O_2 + H_2O

or more simply:

carbon + hydrogen + sunlight ----> oxygen + water + energy

Full-scale photosynthesis probably evolved in steps over several hundred million years. The first forms of photosynthesis were much simpler, probably resembling just PS I or PS II. Hydrogen sulfide, which was present in the early atmosphere, may have been used as a source of hydrogen instead of H_2O. Because oxygen was not produced, these first forms of photosynthesis are called anoxygenic, and there are some bacteria today that still use this low-level form of photosynthesis.

We know that by about 3.5 billion years ago the cyanobacteria were using full-scale photosynthesis because they left behind fossilized remnants. Cyanobacteria were probably the dominant form of life at this time and lived in huge shallow-water colonies where sunlight was abundant. They formed mats on the sea floor that would eventually become covered with limestone and other sediments, cutting off the sunlight, causing the organisms to die. Then a new bacterial mat would form on top of the dead one until it too was covered with sediment. Layer upon layer, sometimes millions of them, eventually hardened into rock. These are called *stromatolites* and we find them today at many locations around the world. Some are dated to 3.5 billion years old, making them the world's oldest fossils of life. The oldest stromatolites show that cyanobacteria were widespread by 3.5 billion years ago, and that life had invented photosynthesis by this time.

The advent of photosynthesis eventually triggered one of the most drastic changes in the history of life on Earth—the *great oxygen catastrophe*. Over the next billion years the atmosphere was transformed from having almost no oxygen to having abundant oxygen. However early life evolved in an oxygen-free environment and oxygen was deadly for it. As photosynthesis became widespread, life was poisoning itself, and this became the first great survival crisis for life on Earth. Was this to be the end of the experiment, or could life adapt to a radically different atmosphere and ocean? Was evolution a powerful enough force to overcome the crisis? We know that life did survive, and much more. Out of the great oxygen crisis came a new form of life that exploded in complexity. This was *Eukarya*, the subject of the next chapter.

The Sea of DNA

Geocosmic Time: Day Three, 5:30 am to 10:50 am

For a billion years Earth was dominated by the prokaryotes. The two broad types, bacteria and archeans, were still very primitive. This does not seem like an exciting time in our story—a billion years of slime and crust—but it was an important time of experimentation and innovation. In addition to photosynthesis, many new energy systems were developed – some were failures and some are still in use today; and new internal structures evolved as the fundamental machinery of life was gradually being perfected. Yet the pace of change was slow: if you sampled life 3.5 billion years ago and then a billion years later, there would not seem to be much difference. This is especially striking when compared to the last *half* billion years (500 million years) before today, when life changed from microscopic organisms into large and magnificent forms like trees and humans.

Still, as unexciting as prokaryotes are, they are the most ancient and versatile form of life; they survived every crisis and catastrophe imaginable. Their staying power is probably because of two things they do very well. First, they can reproduce at staggering rates, making nearly perfect copies of themselves; while it takes a human nine months to produce offspring, bacteria can duplicate themselves in minutes. Second, prokaryotes can exchange genetic material among themselves fairly easily—we call this *horizontal gene transfer*. This is quite different from the more familiar *vertical* gene transfer in higher organisms where genes move from parents to offspring.

Several mechanisms of horizontal gene transfer have been discovered since the 1950s by biologists studying bacteria. In some cases two prokaryotic cells come into contact and exchange sections of DNA called *plasmids*. Plasmids can move from one cell to another carrying copies of sections of DNA. When a plasmid arrives at the recipient cell it brings the source DNA to the DNA of the recipient cell. The two interact and the genome is modified and expanded.

Another mechanism of horizontal gene transfer involves *viruses*, which are essentially chunks of DNA or RNA surrounded by a protective protein envelope. They can move around inside of a cell and seem to be alive, but most biologists do not consider viruses to be a form of life because they are parasitic and cannot live without a host. They can be found in all living things, and those that live in bacteria are often called bacteriophages or simply *phages*. Phages can make copies of a section of DNA in one cell, then leave that cell and enter another cell.

It is believed that early prokaryotes of many kinds shared genes easily in this way, so that life in the ancient oceans was almost like a communal sea of DNA. The best innovations—those that enhanced survival—spread quickly, and this is why the prokaryotes have survived longer than any other form of life and have adapted to every possible environment on Earth. This also accounts for how readily bacteria can become resistant to antibiotic medicines today. If just one of the millions of bacteria exposed to an antibiotic drug survives, it can spread the DNA sequence that lies behind that resistance, and a large number of bacteria can soon become resistant to the drug.

If survival advantages spread so quickly during the age of bacteria, why was the pace of change so slow? Probably because the whole strategy of the prokaryotes was to simply copy over and over the little things that enhanced survival, while innovation and radical new approaches were rare. As we will see in Chapter Six, it was sexual reproduction, invented by a new form of life called *eukarya*, that brought widespread innovation and caused the pace of evolution to skyrocket.

The molecular and geological evidence we now have shows a clear progression from simple bacterial life to larger and more complex organisms over billions of years and we can now understand the mechanisms of this progression in terms of DNA and RNA. Yet the ideas of evolution and the interrelatedness of all life were put forward in the 1800s long before there was any knowledge of DNA.

Science and Discovery

Biological Evolution

Until the mid-1800s it was widely believed in the western world that humans, and other higher life forms, were created by God in a short time not long ago, and that lower life forms such as insects, worms, and mold could arise *spontaneously* in swamps and other places of decay. These ideas were not seriously challenged until biology and chemistry matured in the mid-1800s.

The year 1859 was pivotal in setting the stage for our modern understanding. In that year, Louis Pasteur discredited the theory of spontaneous generation in an experiment before the French Academy of Sciences, showing that microorganisms arose from parents resembling themselves. This idea helped solve the mystery of diseases like rabies, anthrax, and cholera, and led to the first vaccines and even an understanding of the fermentation process used in wine-making.

Also in 1859 Charles Darwin published his book *On the Origin of Species by Means of Natural Selection, or the Preservation of Favored Races in the Struggle for Life*. This unwieldy title is today shortened to *The Origin of Species*. Although most people associate Darwin with the theory of biological evolution, he was not the first to propose this idea—ancient Greeks, like Anaximander, suggested that man descended from lower animals. Nor did Darwin ever use the word "evolution" in his 1859 book. He was, however, the first person to recognize the relatedness of all life, and the major driving mechanism behind biological evolution called natural selection.

The Origin of Species was the result of almost thirty years of work, starting with the five years Darwin spent as a naturalist on board the *HMS Beagle* exploring South America and the Galapagos Islands. After extensive study of both fossils and living things, Darwin proposed that all life on Earth is related and descended from a common ancestor, and that complex creatures evolved naturally from simpler ones. He identified the mechanism of evolution as *descent with modification* followed by *natural selection*. By descent with modification, Darwin meant that living things pass on their traits to their offspring with slight differences that he attributed to random change. Today we understand these differences to be the result of the occasional copying errors in the DNA replication process, known as *mutations*. In higher organisms, the complex mixing of genes from sexual reproduction brings about even more change in the offspring.

Darwin's idea of natural selection was that if a newborn individual had a certain trait that enhanced survival, it would be more likely to reproduce and pass along that trait to its offspring. Individuals without this advantageous trait would be less likely to survive and have offspring. Over many generations, favorable traits would become more common and unfavorable traits less common—a favorable trait being one that enhances survival. Darwin recognized survival to produce offspring as the primary driver of evolution. Today this is popularly referred to as "survival of the fittest." We can sum up Darwin's two ideas about evolution as follows:

- All living things produce offspring that are slightly different from themselves. If offspring were always perfect copies of parents, there would be no evolution!
- Any of the slight differences in offspring that enhance survival are more likely to be passed on to the next generation. Surviving to produce offspring is all that matters!

There is no question that these two things—descent with modification and natural selection—drive evolution. A classic example of natural selection was observed during Darwin's time near Manchester, England, where the peppered moth lives. It had light-colored wings that camouflaged it from predators when it rested on the white bark of birch trees. In the late1800s, heavy industry in Manchester increased, and the soot from coal burning turned the bark of the birch trees nearly black. Amazingly, the moths adapted to this change by developing dark-colored wings that still camouflaged them against the dark birch bark. How did they do this?

While the birch bark was still white, it was known that about 2% of these moths were born with dark-colored wings, and 98% were born with light-colored wings. While the birch bark was still light-colored, the dark-colored moths struggled to reproduce because they were easily seen and eaten by predators. Then, when the birch bark began to darken, these few moths were now camouflaged and were more likely to survive and produce offspring, while the light-colored moths were selected against—they were now more likely to be seen and eaten by predators. The dark-colored moths now had the survival advantage, so they were more likely to survive to produce more offspring with dark-colored wings. In this way the moths adapted to a change in the environment by the process of natural selection, and it happened in a very short time.

A modern example of evolution by natural selection is the use of antibiotic medicines like penicillin. When penicillin was introduced in 1941, it very effectively killed the *Staphylococcus* bacterium, which causes pneumonia and other diseases. Then just a few years later some types of *Staphylococcus* became resistant to penicillin. A few of the bacteria had mutated in a way that allowed them to survive the presence of penicillin. These surviving bacteria reproduced to make new bacteria that were also resistant to penicillin, while others did not survive to reproduce. The bacteria had adapted to an environmental stress (the penicillin) and evolved into a new form that was better able to survive in an environment containing the antibiotic.

Darwin believed that evolution by natural selection happened in small steps. Today this is referred to as *microevolution*, and it has been observed repeatedly. However, in some cases the fossil record in the last half-billion years seems to show dramatic evolutionary leaps that hardly seem to be cases of microevolution. In 1972

Stephen Gould and Niles Eldridge proposed a theory called "punctuated equilibrium" asserting that biological evolution is characterized by sudden leaps in development followed by long periods with little change, hardly Darwin's view. Others, such as Simon Conway Morris, take issue with this theory and believe that Gould misinterpreted the fossil evidence. This debate will be revisited in more detail in the next chapter. A larger question about biological evolution that a few scientists have raised is whether the processes of mutation and natural selection *alone* can account for the progression from primitive bacteria billions of years ago to modern humans today. Could there be additional, undiscovered mechanisms that drive evolution?

Darwin was one of the first scientists to understand evolution as both the organizing theme behind life on Earth and the process that drives life toward greater complexity. His theory goes far to explain how life evolves from previous life, but it does not explain how life first began. In the final paragraph of *The Origin of Species* he says "the Creator originally breathed life into a few forms, or into one." He may have been yielding to religious views of the day because he wrote privately that life probably arose through chemistry "in some warm little pond, with all sorts of ammonia and phosphoric salts, light, heat, electricity, etc., present."

No one can yet fully explain how life got started on Earth. And equally remarkable is that once life got started, it then had almost 4 billion years of relative safety so that it could keep evolving into more complex forms. Life may have gotten started in many locations throughout the universe, but how many of these remained hospitable for billions of years so that complex higher life could evolve and eventually become intelligent, self-aware beings? As yet, we do not know.

Part Two Summary

Geocosmic Time: Day Three, 12:00 am to 10:50 am

By about 4 billion years ago the molten inferno that was early Earth began to cool enough for oceans of liquid water to form and cover our planet. Solid crustal plates formed under the oceans, and some would later thicken into continents. Somewhere in the oceans, perhaps deep down near hot water vents, a remarkable molecule somehow assembled. It could store information in its chain-like structure, it could use that information to build proteins by connecting chains of amino acids, and it could relentlessly make near-perfect copies of itself. Such a molecule became the first life, and it probably grew in complexity and sophistication to become the DNA/RNA complex that is at the heart of all living things today.

At some point the DNA/RNA became enclosed in a membrane (perhaps made of proteins that it built) to form the first cells, and life would remain as single cells for another 3 billion years or so. We call these first living organisms prokaryotes and they were much like today's bacteria, with little internal structure and pieces of DNA floating freely. This was a highly effective model that arguably produced the most successful and resilient form of life in Earth's entire history. The bacteria have withstood an endless barrage of catastrophes and crises for the last 3.8 billion years and still comprise 70% of the living biomass on Earth.

Some primitive bacteria migrated to the surface of the oceans and learned to acquire energy directly from the Sun by developing photosynthesis. Although photosynthesis produced abundant energy for the organism, it also released oxygen, which was a problem: the prokaryotes had evolved in a carbon dioxide atmosphere, and oxygen was poisonous for them. Because of photosynthesizing bacteria, Earth's atmosphere eventually changed from carbon dioxide-dominant into one containing large amounts of oxygen. Although this took a billion years, it finally became a crisis for prokaryotic life. Large numbers of prokaryotic species probably went extinct in this crisis, but some survived. The prokaryotes that survived had either retreated deep underground to oxygen-free environments where they could continue their anaerobic lifestyle, or they adapted to the oxygen-rich environment near the surface of the Earth by inventing a whole new way to acquire energy based on breathing oxygen. This was the start of a new era for life on Earth, the beginning of complex life. From crisis came great opportunity.

Part Three

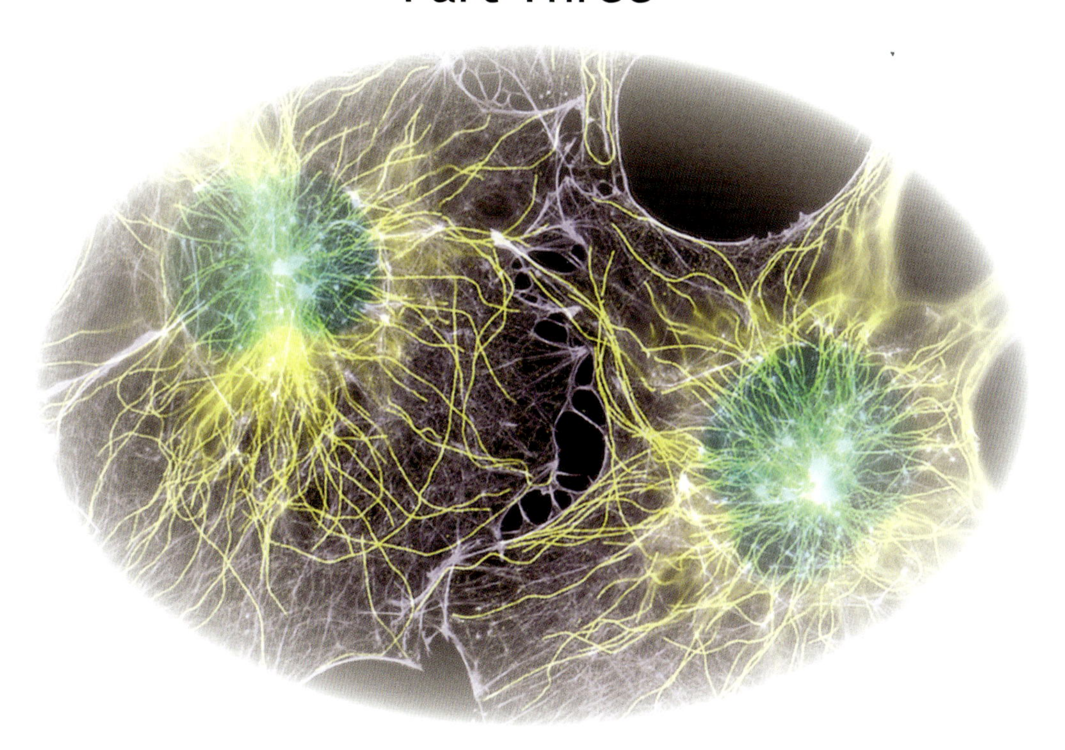

The Age of Complexity

PART THREE

THE AGE OF COMPLEXITY
Life Goes On (and On)

All of life on Earth was prokaryotic for its first billion years or so. This was the age of bacteria. Then, about 2.5 billion years ago, the greatest survival crisis in the history of Earth began to unfold: the atmosphere became deadly poisonous for life, and the poison was oxygen. The *great oxygen catastrophe* by itself might have spelled the end of life on Earth, but to make things even worse the planet plunged into a massive deep-freeze for several hundred million years—the *snowball Earth* event. Life was pushed to the brink of elimination by this double crisis, yet somehow it survived, and the result was a new and much more sophisticated form of life. This was the beginning of the great domain of life we call *Eukarya*, an entirely new pathway that led to the complex forms that are now so prevalent on the Earth—everything from petunias and squids, to horses and humans. The evolutionary leap from prokaryotic life to eukaryotic life was one of the most significant events in the history of Earth.

The Time Conext for Part Three

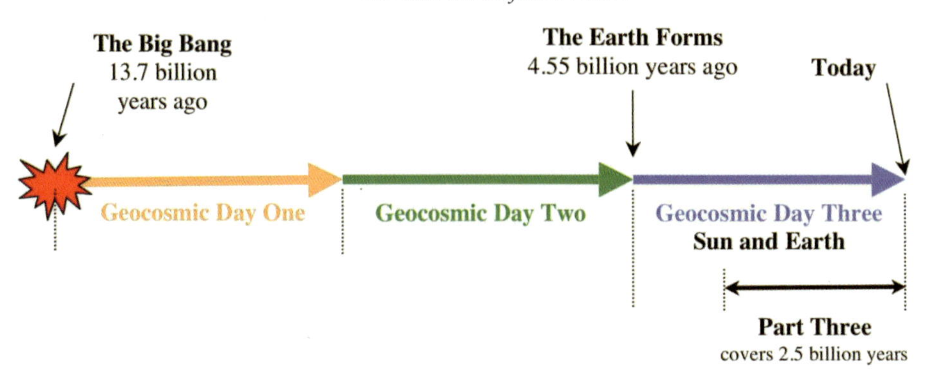

Main Events of Part Three: The Age of Complexity

All dates are approximate

Event	Years Ago	Geocosmic Time
		(all are Day 3)
The great oxygen catastrophe	2.45 billion	11:00 am
First Eukaryotes	2.5-2.3 billion	11:00 am—noon
1st snowball Earth period	2.3-2.1 billion	Noon – 1:00 pm
Plants, animals, and fungi diverge	1.5-1.2 billion	4:00 pm – 5:40 pm
First multi-cellular life	2.0-1.5 billion	1:30 pm – 5:40 pm
2nd snowball Earth period	775-630 million	7:50 pm – 8:50 pm
Cambrian Explosion	540-500 million	9:10 pm – 9:20 pm
First fishes (and vertebrates)	510 million	9:20 pm
First life on land (bryophytes)	475 million	9:30 pm
First land animals (bug-like arthropods)	415 million	9:50 pm
First amphibians	350 million	10:12 pm
First reptiles	320 million	10:18 pm
First mammals	248 million	10:42 pm
First dinosaurs	230 million	10:48 pm
First birds	160 million	11:10 pm
Dinosaur extinction	65 million	11:39 pm
Mammalian radiation	60 million	11:41 pm
First primates	55 million	11:43 pm
First apes	20-15 million	11:54-11:55 pm
First gorillas	9-8 million	11:57 pm
The Hominid lineage diverges	7-6 million	11:58 pm
-------- Looking Ahead -------		
Genus *Homo* appears	2.5 million	11:59 pm
First *Homo sapiens*	200,000	11:59:56 pm
First civilization	5,200	11:59:59.9 pm

Chapter 6
Eukarya: The Power of Cooperation

The Great Oxygen Crisis

Geocosmic Time: Day Three 11:00 am

Earth's early atmosphere should have filled with oxygen after the cyanobacteria began churning out vast quantities of O_2 as a by-product of photosynthesis about 3.5 billion years ago. However fossil evidence shows that there was a one billion-year time lag from when the cyanobacteria first began producing oxygen to the time it showed up in the atmosphere. It should not have taken nearly that long to fill the atmosphere, so why *did* it take so long?

The answer is that the oxygen being produced by photosynthesis was going somewhere else instead of into the atmosphere. It was combining with iron in the ocean and settling out as rust (iron oxide). This can be seen clearly today as reddish streaks in sedimentary rocks called *red beds* or *banded iron* (Figure 6-1).

Figure 6-1

Cross section of rock from the Archean banded iron rock formation in western Australia. The metamorphosed layers of shale and chert were originally deposited in an ancient ocean environment.

Courtesy: Eric D. Miller Collection

Banded iron is most common in rocks dating from about 3.0 to 2.5 billion years ago, and this is the direct evidence that large amounts of oxygen were being produced at this time and then being soaked up by the iron in the ocean. Eventually, by about 2.5 billion years ago, most of the iron in the oceans had combined with oxygen and settled out, and finally the oxygen was free to enter the atmosphere. In a relatively short time (perhaps ten or twenty million years) the atmosphere changed drastically as oxygen levels climbed as high as 20%. This sudden rise in atmospheric oxygen—the *great oxidation catastrophe*—has now been dated to about 2.45 billion years ago (Kump 2008). Along with this rise in oxygen was a drop in carbon dioxide, also caused by the photosynthesis of the cyanobacteria; and less CO_2 in the atmosphere usually means Earth will cool.

Early life was anaerobic; that is, it lived without oxygen. Oxygen was poison for the early prokaryotes, so the increasing levels of oxygen became a survival crisis. Life was poisoning itself with the oxygen it produced from photosynthesis, and most of the life on Earth was probably not able to survive. Some species were able to retreat into oxygen-free environments where they still live today—prokaryotes have been recently discovered deep (miles) underground, still living an oxygen-free lifestyle.

However some prokaryotes apparently learned to protect themselves from the harmful effects of oxygen, and began to use it in a new energy system, *aerobic respiration*. This process entailed breaking down food molecules (such as sugars made by photosynthetic organisms) to derive hydrogen and carbon, then combining these with oxygen to produce about ten times as much ATP (fuel) as anaerobic respiration could. Some of the hydrogen that was stripped off of food molecules could combine with the toxic oxygen to make harmless water. A simplified version of aerobic respiration can be written as:

carbon and hydrogen (from food) + oxygen ----> carbon dioxide + water + energy

The prokaryotes that invented aerobic respiration had found a solution to the oxygen crisis—they could not only protect themselves from the poisonous effects of oxygen, but they could also produce much more energy by using oxygen.

However, the oxygen problem was only one part of the crisis that life on Earth faced 2.4 billion years ago. The drop in atmospheric CO_2 caused the Earth to cool, triggering the second catastrophe: a global deep-freeze that has come to be called the snowball Earth event. This was the reverse of our problem today with global warming caused by a rise in CO_2. We now know that Earth's average temperature is closely tied to atmospheric CO_2 levels. When CO_2 levels are high, heat is trapped by the atmosphere and the temperature goes up (the greenhouse effect), and when CO_2 levels are low, the temperature goes down. Contributing to this was the Sun itself: it was not as bright then as it is now.*

By about 2.3 billion years ago, with CO_2 levels falling, Earth slipped into the most massive deep freeze of all time. Earth was covered almost entirely by ice up to a half-mile

* The Sun's luminosity has been steadily increasing as hydrogen in the core is depleted and the core shrinks.

thick; and the white surface of the planet further contributed to the deep freeze by reflecting most of the Sun's energy. Sunlight was cut off from nearly all life, except near the equator; in the cold darkness of the oceans nearly all of life on Earth was probably wiped out. This was Earth's first mass extinction event, when the experiment of life very nearly ended. Yet, at this darkest of times, as the deep freeze was setting in for hundreds of millions of years, a new form of life appeared. These were the eukaryotes.

Science and Discovery

Endosymbiosis and the Rise of Eukarya

The earliest *fossil* evidence of eukaryotic life is from about 1.8 billion years ago, but molecular techniques have pushed this date back to about 2.5 billion years ago (Lipps 2001). The first eukaryotes were single-celled organisms like their prokaryotic ancestors, but much larger and more complex. Eukaryotic cells today are generally about 50 to 100 times bigger than prokaryotic cells, and they contain internal structures, called organelles, enclosed in membranes.

The first organelles were probably devoted to acquiring energy. Some eukaryotes developed an organelle, the *mitochondrion*, where cellular respiration took place, allowing the cell to derive energy from oxygen and sugar. Other eukaryotes developed the *chloroplast* where photosynthesis took place, allowing the cell to derive energy from sunlight and carbon dioxide. Early eukaryotes with mitochondria eventually evolved into oxygen-breathing animals, while those with chloroplasts later became carbon

Figure 6-2

Eukaryotic cells are enormously more complex than prokaryotic cells.

This color-enhanced fluorescence microscope photo shows two animal cells. In yellow are microtubules, in purple is actin, and in green are the cell nucleii.

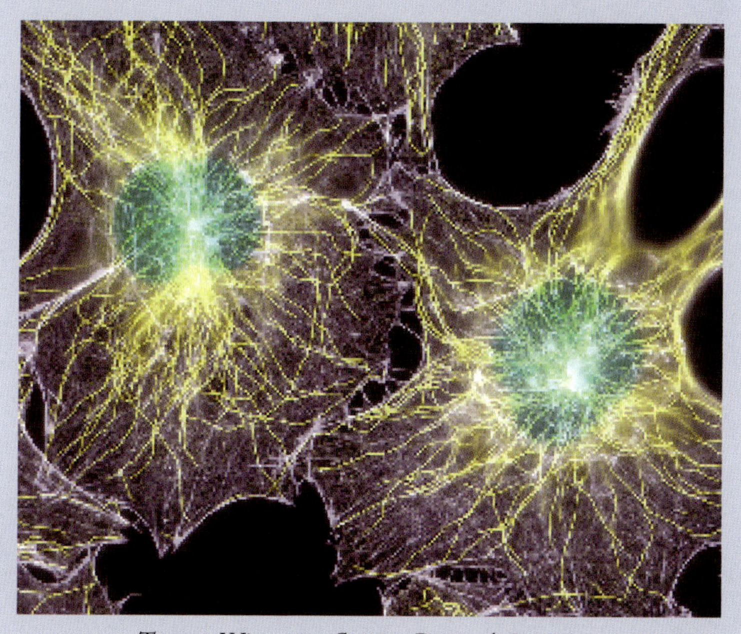

Torsten Wittmann, Scripps Research Institute

dioxide-using plants. Another important organelle that probably evolved somewhat later was the cell nucleus, which held the DNA and organized it into chromosomes. As we will see, this innovation opened the door to an entirely new method of reproduction. How did simple prokaryotic life make such a leap in complexity to the eukaryotic form of life?

Lynn Margulis, working at Boston University, provided an answer to this question in 1967. The idea she proposed, called the *endosymbiotic theory*, met with considerable resistance at first, but is now widely accepted. Margolis noticed that the organelles of eukaryotes looked like bacteria. She suggested that large prokaryotes engulfed smaller prokaryotes (a process called endocytosis) to form a mutually beneficial (or symbiotic) relationship. Margulis theorized that mitochondria were originally prokaryotes that had developed oxygen respiration. The larger prokaryotes that swallowed them up were probably struggling to survive in the oxygen-filled environment and the smaller oxygen-breathing organism could supply energy for the larger organism. The larger organism now had an energy supply in an oxygen-rich environment, while the smaller one benefited by being protected and supplied with nutrients. This symbiotic relationship was a win-win situation that launched the eukaryotic revolution.

The development of mitochondrial structures in early eukaryotic cells was the first step toward animal life. While animals use mitochondria and oxygen respiration for their power source, plants use chloroplasts and photosynthesis to supply their energy. The first chloroplasts probably evolved when a large prokaryote engulfed a smaller photosynthesizing prokaryote. Again, a mutually beneficial arrangement gave rise to a more complex and, better adapted, form of life. These two new forms of eukaryotic life—one with mitochondria where oxygen respiration took place and the other with chloroplasts where photosynthesis took place—launched the remarkable relationship between plants and animal, with their exchange of oxygen and carbon dioxide.

When Margulis first proposed this scenario, with one organism engulfing another, there was no real evidence to support or refute her idea. However, good theories in science make predictions that can be tested, and her theory did just that. She predicted that if eukaryotic organelles were in fact engulfed bacteria, there should be DNA in these organelles that resembled bacterial DNA. With advances in technology in the 1980s, bacterial DNA was found in the mitochondria and chloroplasts, lending strong support for the Endosymbiotic Theory. Further support came with the finding that mitochondria and chloroplasts replicate by binary fission as bacteria do, and that they have ribosomes resembling those of bacteria.

Since the late 1960s when the endosymbiotic theory originated, the archeans have been identified as the other major domain of life that was also prokaryotic. Some scientists now believe that it was the merging of archeans with bacteria that gave rise to the domain Eukarya. The symbiosis that gave rise to the eukaryotes was the first of many collaborative efforts that advanced life toward greater complexity. Even today, in the era of humans, collaboration and cooperation have proven to be very powerful strategies.

Snowball Earth

As life evolved on Earth over the last 4 billion years, it faced monumental hardships, including meteor and comet impacts, wild climatic swings, and a radical change in the composition of the atmosphere. Yet the most harsh and long-lasting challenge was probably the first snowball Earth event that nearly eliminated all of life.

Geologists have found evidence of this event in the form of glacial deposits known as *tillites*. Tillites are common in the far northern and southern latitudes, where glaciers have been active throughout Earth's history, but unusually thick tillite deposits have been found that suggest much longer and deeper periods of glaciation. Because these deposits have been found worldwide and at latitudes close to the equator, it appears that the Earth was covered almost entirely by glaciers, with ice as thick as a half-mile for a hundred million years or more.

In 1992 Joseph Kirschvink from Cal Tech identified this super ice age and gave it the name *snowball Earth*. We have now found additional tillite deposits from a later time indicating that there were actually two snowball Earth periods. The two snowball periods have been dated to roughly 2.3 to 2.1 billion years ago, and the more recent period to about 600 to 800 million years ago. The more recent period was actually a series of three deep freezes, each lasting about 10 million years. During both snowball periods, life still resided only in the oceans, but the oceans were frozen to depths of a half-mile or more, blocking out sunlight completely. Any open water near the equator would have been critical for the survival of life. Earth was brutally cold during the snowball times, with surface temperatures well below zero Fahrenheit for tens of millions of years.

The dates of the first and second snowball periods coincide with some of the most important periods in the history of life on Earth: the first is near the time when eukaryotes first appeared, while the second snowball period was followed by the *Cambrian explosion* of complex animal life which we will explore later in this chapter. Until recently it was believed that eukaryotic life emerged just after the first snowball period, perhaps 2 billion years ago, but newer molecular evidence suggests that the eukaryotes probably evolved as the massive freeze-up was setting in, closer to 2.5 billion years ago (Lipps 2001). Perhaps endosymbiosis, in which prokaryotes merged to form eukaryotic cells, was the survival advantage that allowed life to withstand the extreme conditions.

In addition to the major snowball periods, Earth has experienced many smaller cold periods, the so-called ice ages. The most recent ice age ended about 12,000 years ago, setting the stage for human civilization. These ice ages, however, were mild compared to the snowball events. During recent ice ages, glaciers covered most of Canada and Northern Europe, but the mid-latitudes remained ice-free.

When ice builds up on land, the sea level drops, exposing land bridges like the one through the Bering Straights that connects Siberia to Alaska. Apparently early modern humans used this crossing to migrate from Asia to the Americas during the last ice age (Chapter Eleven). As Earth has warmed in modern times and the polar ice sheets continue to melt, sea levels have risen. By the year 2000 most climate scientists agreed that the well-documented rise in atmospheric CO_2 during the last few decades has caused a rise in the average temperature of the Earth. Despite the widespread consensus in the scientific community, however, a survey conducted in 2009 by the Pew Research Center found a sharp decline in the number of Americans who say there is solid evidence that global temperatures are rising (Pew Research Center 2009).

Scientists are still trying to develop a complete understanding of climate change, but there is widespread consensus that the main driver of global warming is one thing: the greenhouse effect. Certain gases in the atmosphere, like carbon dioxide, methane, and water vapor, are called *greenhouse gases* because they trap heat in the same way a greenhouse warms up. Because the most abundant greenhouse gas is carbon dioxide, we can say quite simply that Earth's temperature is affected more that anything else by the level of carbon dioxide in the atmosphere. When Earth's atmosphere is high in CO_2, as it was for its first 1.5 billion years, Earth is warm. When CO_2 levels drop, so does Earth's temperature.

The first snowball event, starting about 2.3 billion years ago, was most likely triggered by the drop in CO_2 levels brought about by the photosynthesis of cyanobacteria. The cyanobacteria pulled CO_2 out of the atmosphere and replaced it with O_2, which is not a greenhouse gas. Other factors may have contributed to the snowball event, including variations in the sun's intensity, variations in Earth's orbit, and the wobble in the Earth's axis of rotation.

Snowball and ice age events should have the tendency to run out of control, leaving the Earth permanently frozen. As cooling sets in, more and more of Earth's surface gets covered by snow, which reflects sunlight. While bare ground reflects only about 20% of the sun's energy, snow and ice reflect about 80%. As Earth gets covered by more snow, more of the sun's energy gets reflected, which causes a further drop in temperature, producing more snow coverage, and so on, leading to a "runaway icehouse." Similarly, during times of warming, Earth should tend toward a "runaway hothouse." As Earth's temperature rises, more water evaporates into the atmosphere and this water vapor is a greenhouse gas that traps heat. The temperature rises further, driving more water vapor into the atmosphere, causing even more heating, and so on. Venus, where surface temperatures reach 800 degrees, is the victim of a runaway greenhouse effect, but Earth somehow manages to pulls out of these extremes. It is remarkable that the temperature of the Earth has remained for billions of years within the relatively narrow range where water is a liquid. How does the Earth do this?

Science and Discovery

The Global Thermostat and Plate Tectonics

In 1981 the global thermostat mechanism was first recognized by geochemists James Walker, Paul Hays, and James Kasting. This mechanism regulates Earth's temperature by cycling carbon in and out of the atmosphere as CO_2. When things are getting too hot, CO_2 is pulled out of the atmosphere, which causes a cooling, and when things get too cold, atmospheric CO_2 levels increase. At the heart of this thermostat mechanism is weathering, which is the set of processes that break rocks and boulders down to sand. Water, wind, and freeze-thaw cycles are the main players in weathering. Granite and other types of rock contain silicate minerals like mica and feldspar, and these silicates combine with CO_2 in the atmosphere to form limestone. The important result is that CO_2 is removed from the atmosphere. This process is described in the following chemical reaction:

$$CaSiO_3 + CO_2 \text{ ----> } CaCO_3 + SiO_2$$

Or more simply,

silicates + carbon dioxide -----> limestone + sand

This means that $CaSiO_3$, the silicate produced by the weathering of rocks, combines with carbon dioxide from the atmosphere to produce limestone and sand, which get deposited in the ocean as sediments. This pulls CO_2 out of the atmosphere and buries it under the ocean. However, this chemical reaction is temperature dependent, and it runs faster at higher temperatures. So, when the Earth's temperature is high, more CO_2 is pulled out of the atmosphere, which causes things to cool. When Earth is cooling off, this reaction slows down, pulling less CO_2 out of the atmosphere. Combined with the effects of volcanism, which constantly pumps CO_2 into the atmosphere, the greenhouse warming will increase, and the temperature will go up. This remarkable thermostat mechanism has kept Earth's temperature within the narrow range of about 20° to 40°C required by complex animals and plants. Since the end of the last snowball event about 600 million years ago, the global thermostat has prevented a runaway icehouse or hothouse and made complex life possible.

The weathering processes at the heart of the global thermostat are possible because new rocks are constantly being exposed to the atmosphere by the churning and uplifting of the Earth's crust. This churning is the result of another Earth specialty: *plate tectonics*. Mountain building and other motions of the crust are due to the fact that Earth's crust is made of large granitic plates that float on denser basaltic magma below them. Some of these plates thickened to become continents. Alfred Wagener first proposed the theory of plate tectonics in 1912 when he suggested that continents float and move slowly. He noticed that the east coast of Brazil seems to fit with the west coast of Africa, and he found similar rock types on both coasts, suggesting that they were once connected. Yet Wagener did not live to see this theory accepted. There were few believers until the 1960s when supporting evidence emerged.

We now know that some of the floating plates thickened to become the first continents, rising from the oceans between about 3.0 and 2.5 billion years ago. Deep-sea exploration has revealed a prominent mid-Atlantic ridge where the continents of South America and Africa were once connected, and a large amount of other evidence strongly supports the theory of plate tectonics.

Plate tectonics is apparently unique to the Earth. Close-up images of other planets and moons in our Solar System, sent to us by exploratory spacecraft, show a distinct absence of linear mountain ranges like the Rockies, the Alps, and the Himalayas. Earth's great mountain ranges are the result of colliding plates where one plate is pushed down and the other is pushed up. Other planets and moons have large circular mountains that are volcanic, but nowhere else in the Solar System do we see chains of mountains like those found on the Earth. It seems that Earth is the only body in the solar system with plate tectonics, and therefore the only world with prolific weathering of rocks. This makes the global thermostat mechanism possible, which in turn keeps the global temperature within the narrow range necessary for complex life to survive.

The Sexual Revolution

Geocosmic Time: Day Three, beginning about 11:00 am

The emergence of eukaryotic life some 2.5 billion years ago not only brought new ways to supply energy and better ways to organize the cell, but also a new reproductive strategy that had a major impact on evolution. With the older *asexual* strategy, still used today by bacteria and other prokaryotes, an organism makes a nearly exact copy of itself, and it does this over and over. This maximizes the chances of survival through sheer numbers, and it increases the chance that a beneficial mutation might occur. In the common asexual process called *binary fission*, the disorganized DNA floating inside a bacterial cell replicates, then the two duplicate clumps of DNA move to opposite sides of the cell, and finally a new cell membrane grows in-between, splitting one cell into two nearly identical cells.

Bacteria are constantly doing this and can increase their populations at staggering rates. The disadvantage of this strategy is that there is very little change from generation to generation. This is fine as long as the environment of the organism remains unchanged. However if the environment changes significantly, as it inevitably does, these organisms have limited ability to adapt, and they may not survive. For the prokaryotes, evolutionary advances depend on mutations when the DNA replicates, and the large numbers of new organisms that are constantly being produced.

The strategy of sexual reproduction, developed by the eukaryotes, brought rich diversity to each new generation and a much greater ability to adapt to changing environments. It is not known how eukaryotes developed sexual reproduction, but it undoubtedly happened in steps over at least a billion years. Today if we study protists (one-celled eukaryotes like the amoeba or paramecium) we see many forms of sexual reproduction that probably represent steps toward the complex process that higher plants and animals use. Once it got started, sexual reproduction drastically changed the game of survival and evolution.

In a eukaryotic cell, the genome is much more organized than in a prokaryotic cell. In a eukaryote, the DNA is enclosed in the cell nucleus, and within the nucleus it is separated into chromosomes. The entire human genome is contained in 23 chromosomes, but it is also backed up in a duplicate set of chromosomes, making a total of

46 chromosomes in every human cell (except the egg and sperm cells). Cells with this double set of chromosomes are called *diploid,* or *2n,* while cells with only a single set of chromosomes are called *haploid,* or *n.*

The number of chromosomes varies across the spectrum of eukaryotes. While humans have 23 pairs of chromosomes, cats have 19 chromosome pairs (or 38 total), goldfish have 40/80, and corn has 10/20. There are advantages to having two sets of instructions in our cells because sections of DNA can become damaged or changed. However the two sets of chromosomes are not exact duplicates, and this becomes important in sexual reproduction.

Each chromosome is a long chain of DNA, and shorter sections of that chain encode specific information. These shorter sections of DNA are called genes, and a typical chromosome may contain about 5000 genes. In eukaryotes the entire genome is organized somewhat like an encyclopedia with many volumes. While we could write an encyclopedia with all the information in one very thick volume, it would be unwieldy and difficult to use. So we break up the information into smaller volumes. Then within each volume there are specific topics, each topic is described in sentences, each sentence is made of words, and each word is made up of individual letters. The genetic parallel to this is shown in Figure 6-3.

Figure 6-3
Genetic information is organized in eukaryotes like an encyclopedia with many volumes.

DNA	Encyclopedia
Single Nucleotide (one rung of the DNA ladder)	Single letter
Nucleotide triplet (also called a codon)	Word
String of Nucleotide triplets	Sentence (strings of words)
Gene	Topic
Chromosome (many genes)	Volume (many topics)
Genome (many chromosomes)	Encyclopedia (many volumes)

The sexual revolution of the eukaryotes was much more than just a better way to organize the genetic information. Sexual reproduction shuffles the genes of each parent, then combines the shuffled genes to produce offspring that are significantly different from the parents. In contrast, the asexual reproduction of prokaryotes creates offspring that are nearly exact copies of the parents.

Sexual reproduction is based on the fact that eukaryotes can make a specialized kind of cell with just a single set of chromosomes rather than the usual pair. This haploid cell is the gamete, or sex cell. Gametes are manufactured in the process called meiosis. In humans, this happens in either the ovaries, where egg cells are made, or in the testes where sperm cells are made; and of course this brings up the two sexes: the eukaryotes invented male and female. The process of meiosis begins with a diploid cell containing the usual double set of chromosomes in its nucleus (see Figure 6-4). In the first step, the chromosomes replicate almost exactly, then in an intricate dance, corresponding sections of chromosomes are traded, and finally these shuffled chromosomes are carefully segregated and sorted into four complete sets. The original cell has turned into four gametes, each with one set of chromosomes. However none of the sets of chromosomes is identical to the original.

Figure 6-4

A comparison of mitosis and meiosis. The end result of mitosis (left) is two diploid cells identical to the original, while the final product for meiosis (right) is four haploid gametes (egg or sperm cells) with a mixture of chromosomes from the original diploid cell.

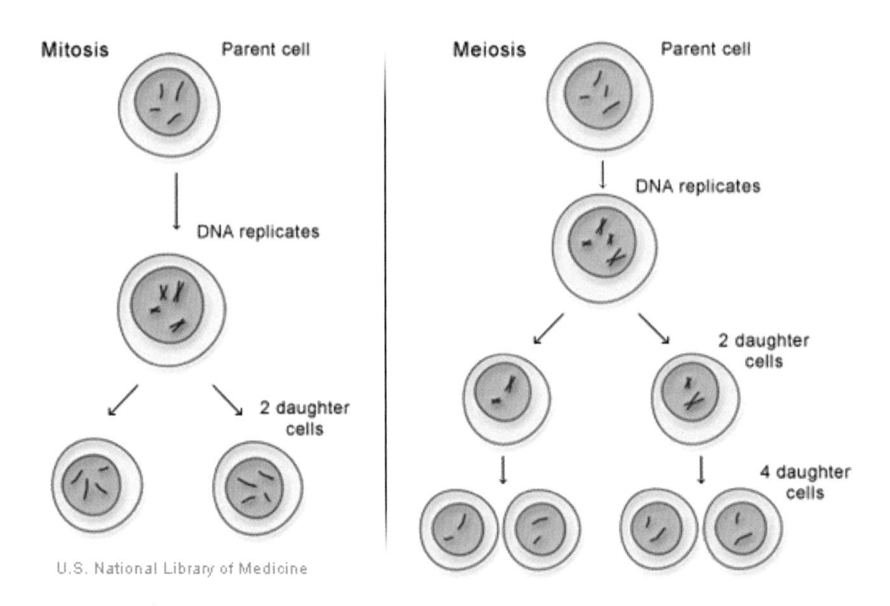

Through the act of sex itself, an egg cell and a sperm cell from different parents come together in some kind of mating process to combine their single sets of chromosomes (in flowering plants, pollen carries the sperm cell). The resulting zygote, or fertilized egg, will now have a full double set of chromosomes with a mix of genes from both parents. Although half of the genetic material came from each parent, the offspring will be something significantly different from the parents.

The single fertilized egg cell now begins to grow by dividing over and over in a process called *mitosis*. In mitosis the double set of chromosomes is duplicated and distributed as one cell splits into two. Through repeated mitotic cell divisions, the fertilized egg cell grows into a full complex organism, all according to the instructions contained in the DNA of every cell. Encoded in that DNA is every detail of how to build and maintain the organism.

The development of sexual reproduction by the eukaryotes was a major turning point in the story of life on Earth. For billions of years, life existed only as one-celled organisms in the oceans, but after sexual reproduction was perfected, the pace of evolution accelerated, leading to an explosion in the diversity and complexity of life forms. There are two main reasons that sexual reproduction had such an impact on the evolution of life. The first is the genetic mixing that comes from sexual reproduction. When gametes (the sperm or egg) are made, the process of meiosis shuffles the genes of each parent. Then, when gametes from two parents combine, there is a further shuffling of the original genetic material. Each new generation is an experiment in life that is put out into the changing environment. The unsuccessful experiments struggle or die and thus will not be likely to pass on their genes, while the offspring that are better adapted to their environment will live to reproduce and pass on their better adapted genes. Each new generation has the potential to be better adapted to the environment than their parents. Every new generation can be an evolutionary step forward. For the prokaryotes, like bacteria, little progress is made with each new generation because the offspring are almost exactly like their parents.

The second reason sexual reproduction had such an impact on evolution is mate selection, or what biologists call *sexual selection*. If the joining of the male and female gametes were purely random, there would still be significant diversity in the offspring. However, mating, especially in higher animals, is not random! Two animals somehow choose a mate and the new organism they produce is very much influenced by that choice. Darwin recognized two kinds of sexual selection in the animal world when he wrote in 1871:

> The sexual struggle is of two kinds: in the one it is between the individuals of the same sex, generally the males, in order to drive away or kill their rivals, the females remaining passive; while in the other, the struggle is likewise between the individuals of the same sex, in order to excite or charm those of the opposite sex, generally the females, which no longer remain passive, but select the more agreeable partners. (Darwin 1871)

Throughout the world of the eukaryotes, sexual selection operates in many ways. The male Bighorn Sheep has developed horns apparently for the sole purpose of battling other males to win the right to mate with the most desired female. Presumably the male that wins is the strongest and will bring the best genes to the female and ultimately produce superior offspring. Other animals use visual displays to attract a mate, like the spectacular tail of the male peacock. And, of course, humans have elaborate courtship and selection rituals. While some women may choose their mate purely on the basis of physical attractions, others may select for intelligence, parenting skills, or other qualities. Clearly the offspring that are produced from sexual selection, and thus the course of evolution, will be influenced by that choice.

In recent times humans have manipulated selection to their advantage, especially in agriculture. Hundreds of years ago the ears of corn grown by Native Americans were only a few inches long. Today's corn, after many generations of selective cross-pollination, produces ears that are typically about 12 inches long. This kind of improvement can be seen in virtually all the foods we grow today, and in all of the domesticated animals we raise. Humans have used selective breeding to alter plants and animals to their advantage, mostly as food sources, for at least 10,000 years. The process of natural selection that drove evolution for billions of years has been altered by "artificial selection." Humans have changed the course of evolution in many ways and it remains to be seen whether this is a blessing, a curse, or some of both.

Although eukaryotic life was a leap in complexity from previous forms of life, it did not replace prokaryotic life—bacteria and archaea are still thriving today. The appearance of the eukaryotes opened the pathway to higher, more complex forms of life, like today's plants and animals. The first eukaryotes were one-celled organisms called protists, like today's paramecium and amoeba, and they flourished for more than a billion years as the highest form of life on Earth. If we could have visited Earth two billion years ago or even one billion years ago when protists ruled, there would still not be much to see—all of life was still single-celled, microscopic, and it lived only in the water (landmasses had risen by this time but were devoid of life). It may seem that not much happened from the time eukaryotes first evolved about 2.5 billion years ago until the first multicellular organisms were proliferating about 1.5 billion years ago. Why was the pace of evolution so slow, especially if sexual reproduction was such an accelerator of evolution?

The answer is that there was an enormous amount of complexity developing within the cell walls of these seemingly simple organisms. Countless new structures and mechanisms developed, such as whip-like tails (flagella) used for propulsion, sophisticated chemical signaling systems, and internal scaffolding structures for transporting large molecules. Huge advances were made in cellular mechanics that later life would build upon, with later single-celled eukaryotes, like yeast or amoebas, reaching a level of complexity akin to a Boeing 747 with six million parts (Vines 2000). From this perspective, it's easier to understand why the evolution of eukarya remained within the membranes of a single cell for so long—there was enormous complexity developing inside these single cells, and it was probably sexual reproduction that made it possible.

Multicellular Life and the Fossil Record

Geocosmic Time: Day Three, starting somewhere between 1:00 pm and 4:00 pm

Sometime between about 2.1 and 1.5 billion years ago, life moved beyond the single cell and began to evolve into simple multicellular forms (El Albani 2010). This was a new collaborative arrangement in which single cells began to join together into larger organisms. Long before this, identical single cells had grouped together in large colonies, but now cells began to take on specialized functions. The ability of cells to differentiate into specialized kinds with specific functions was a significant step toward higher life. In human development, a single fertilized egg divides repeatedly and eventually begins to differentiate into hundreds of different kinds of cells, from liver cells to brain cells, as an embryo matures into a newborn child.

Early multicellular life probably resembled microscopic jellyfish and worms, and it remained at that level until about 600 million years ago. The distinction between plants and animals was also becoming clear by about 1.5 billion years ago. Plants had chloroplasts in their cells where photosynthesis took place; they could capture the Sun's energy directly, while taking in CO_2 and expelling oxygen. Animals had mitochondria in their cells where oxygen respiration took place; they needed food sources such as plants or other animals, and expelled the CO_2 that plants needed; and so the remarkable partnership between plants and animals emerged. Life would have had a very limited future without it.

It is difficult for biologists to know exactly what life was like up until about 600 million years ago because there is almost no evidence surviving today. When soft-bodied organisms died they usually vanished without a trace, or possibly left a microscopic speck in some rock layer that would be almost impossible to find and interpret. However this was about to change. By about 600 million years ago the fossil record begins, when much larger, hard-bodied organisms first appeared. From this time on, the rocks record a clear and relentless progression in complexity that has continued until today. In the last half-billion years complex life as we know it came to be.

Fossils provide almost all of our information about the last 600 million years when life blossomed into complex forms; but fossils are both wonderful and problematic. When any living thing dies, it has the possibility of becoming a fossil. That is very unlikely, however, because when most organisms die, they are eaten by other organisms. Every molecule is stripped off and recycled in some way, leaving no trace. If an organism should happen to die in a sediment, like mud or sand, then be buried immediately and sealed off from oxygen, and if the body parts are hard and become replaced by minerals, then a fossil might form as the sediments harden into rock.

Because fossils form in sediments, they are found mostly in sedimentary rocks (like sandstone and limestone), which are only about 15% of all rocks. The oldest fossils are in metamorphic rocks that were originally sedimentary but were transformed by heat and pressure. These processes, and the buckling and folding that rocks are subjected to, can easily destroy delicate fossils.

The biggest problem with fossils, though, is that someone has to find them. If you would go to the most remote place you could find and bury a precious object as deeply as possible, what would be the chances that someone would ever find it? Like your precious object, most fossils will never be found. All of these problems with fossils lead to one conclusion: they represent a very, very small slice of the history of life. Bill Bryson sums up the problem with fossils in this way:

> Most of what has lived on Earth has left behind no record at all. It has been estimated that one species in ten thousand has made it into the fossil record. Moreover, the fossil record is hopelessly skewed. Most land animals, of course, do not die in sediments. They drop in the open and are eaten or left to rot or weather down to nothing. The fossil record consequently is almost absurdly biased toward marine creatures. About 95% of all the fossils we possess are of animals that once lived under water, mostly in shallow seas. (Bryson 2003)

Nearly all of our knowledge of how complex life evolved comes from fossils, but scientists must fill in huge gaps between very sparse clues. Whole time periods are missing and many transitional forms of life are still undiscovered because we have so few fossils. Some of the first fossils found during the early 1800s were of trilobites. Trilobites lived on the sea floor and resemble the modern horseshoe crab, with a beetle-like appearance. They ranged in size from microscopic to several feet across, and proliferated in the seas for nearly 300 million years until they disappeared in the great Permian extinction event about 245 million years ago. The early fossil record seems to suggest that trilobites appeared rather suddenly, fully formed. These were extremely complex organisms with limbs, gills, antennae, a small brain, and primitive eyes. Trilobite fossils have been found in many locations around the world, as though they appeared simultaneously all over the Earth. This was troublesome for early Darwinians whose theory of evolution rested on small, incremental changes. Some people cited this as proof that a supernatural being had intervened. Today we have dated the rock layers containing the first trilobites to about 543 million years old, but their sudden appearance still seems strange. The trilobites were part of a much larger evolutionary leap that has come to be called the *Cambrian explosion*.

Exploring Deeper

The Cambrian Explosion
Geocosmic Time: Day Three, 9:10 pm to 9:20 pm
In 1909 a paleontologist, Charles Walcott, made a remarkable find high in the Canadian Rockies on an outcropping of shale called the Burgess Ridge. He discovered the richest and most important concentration of fossil animals in history. These animals were living about 505 million years ago in a shallow sea underneath a steep cliff. The cliff collapsed and buried a rich assortment of animal life, preserving their soft

bodies almost perfectly. The result is a snapshot of animal life at this moment and it reveals what seems to be the sudden appearance of many new forms of life.

Walcott spent the next fifteen years, until his death in 1927, collecting and cataloging fossils from Burgess Ridge. He amassed over 65,000 specimens of about 140 different animal species that had never been seen before, but he did not understand the significance of what he had found. He attempted to classify the many strange specimens into a few existing categories, or phyla,* and regarded his discoveries as mere curiosities whose lineages did not survive. His collection sat untouched in a Washington museum from 1927 until 1962 when it was re-examined by Alberto Simonetta, who realized that there was much more to be learned from the Burgess fossils. A team headed by Harry Whittington from Cambridge University returned to the Burgess Ridge and resumed digging and collecting at several quarries. Whittington was the world's authority on trilobites and his two graduate students, Simon Conway Morris and Derek Briggs, would later become stars in the world of paleontology.

Figure 6-5

Anamalocarsis was one of many strange animals that appeared during the Cambrian explosion. In its time this was one of the most deadly predators on Earth, and like many Cambrian experiments, it went extinct long ago.

Courtesy: Peer Ziegler, www.scientificdesign.eu

The new work on the Burgess fossils led Whittington's team to a much different conclusion from Walcott's. They found that these animals represented not just a few previously known phyla, but many phyla that had never been seen before, including almost all of today's. This meant that animal life on Earth had taken a phenomenal leap in diversity and complexity between about 543 million years ago and 505 million years ago when the Burgess fossils were preserved, a leap that soon came to be known as the *Cambrian explosion*.

* The phylum is a broad category at the top of the taxonomic scheme (domain, kingdom, phylum, class, order, family, genus, species) that may be thought of as "body plan." For example the phylum Chordata includes organisms with a notochord, or dorsal nerve tube. In most cases this is a spinal chord, and most members of Chordata are vertebrates, or animals with a backbone-type body plan.

The Burgess Shale and other deposits of the same age from around the world all seem to record the same thing: very few fossils in layers older than about 543 million years, then suddenly a huge number of fossils of many different types, with a high level of complexity and large in size. It seemed that life had remained at the simplest level for billions of years, then rather suddenly, in a few tens of millions of years, all of the basic body designs of today's complex life appeared; the most important was bilateral symmetry, the two-sided body plan that nearly all animals today share. In the context of the three geocosmic days, the Cambrian explosion is a 10-minute flurry at about 9:10 in the evening of the third day, when all of the modern body plans—all of today's phyla—appeared; and many new body plans appeared that eventually failed and no longer exist today. This explosion of life was unparalleled in the innovation and experimentation it reflected. What could have caused it? No one can yet fully explain it.

The mystery of the Cambrian explosion has been compounded by the more recent discovery of a strange kind of animal life that lived just before the explosion, from about 600 to 543 mya. These fossils were originally found in 1946 in the Ediacara Hills of Australia and are now called the *Ediacaran fauna*. They represent a now-extinct kind of animal life resembling giant ferns and sponges with mattress-like quilting and attached to the sea floor. The Ediacarans apparently emerged as the final snowball event was ending and they reflect life's first venture into large-scale forms.

More recently Ediacaran fossils have been found in other locations around the world from the same time period, but no one yet agrees on their relationship to later forms of animal life. Was this a failed first attempt at large-scale life, or did some of the Ediacarans evolve into later forms of animal life that appeared in the Cambrian? Many paloentologists now favor the theory that something caused the Ediacarans to go extinct at about the time of the Cambrian explosion. The large, hard-bodied animals that replaced them in the Cambrian explosion may have been around in microscopic form for millions of years and the demise of the Ediacarans, as well as an increase in atmospheric oxygen, may have cleared the way for an explosion in size. A factor that probably contributed to the failure of the Ediacarans and the success of the Cambrian animals was mobility. The Ediacarans were stuck on the ocean floor and were perhaps smothered by sediments or mats of cyanobacteria, or poisoned by the build-up of toxins. Most of the Cambrian animals were mobile and could move around to find food and more favorable environments. Mobility was a significant advantage for the Cambrian animals that has contributed to the success of animals ever since.

Though the Ediacarans are still an enigma for evolutionary biologists, the fact remains that a worldwide explosion of animal life quickly followed their demise about 543 million years ago. This was the beginning of complex animal and plant life, and from here on, the pace of evolution accelerated spectacularly to produce large and magnificent life forms with ever-greater complexity. However, the cause of the Ediacaran extinction and the Cambrian explosion is still debated, as is the very nature of biological evolution. Does evolution have occasional periods of explosive development, or does it proceed gradually, in continual small steps, as Darwin believed?

Science and Discovery

Contingent or Convergent Evolution?

Chance versus Inevitability: Stephen J. Gould and Simon Conway Morris

The meaning of the Cambrian explosion and the very nature of biological evolution would only become more controversial and heated after Whittington's team re-evaluated the Burgess fossils in the early 1970's. Their initial conclusion was that the fossils came from many new phyla, popularizing the notion of a Cambrian explosion. This became the basis of Stephen J. Gould's 1989 book *Wonderful Life*, one of the most widely read science books ever written. It vividly chronicled the story of the Burgess Shale fossils, with Simon Conway Morris a sort of hero. However Gould went far beyond the story and championed a view of evolution that would become highly contentious. In Gould's view, the Cambrian explosion of diversity and complexity typified the history of evolution on Earth—it proceeded very irregularly with sudden bursts of activity followed by long periods of relative inactivity, a theory called *punctuated equilibrium* that he and Niles Eldridge had proposed in 1973. The vast number of new body plans produced in the Cambrian was later culled down by unpredictable environmental pressures, leaving the forms of animal life we see today. However, which of the new phyla went extinct and which of these survived to become today's creatures, including humans, was a product of chance: it would not come out the same way if it happened again.

At the heart of Gould's view of evolution was the idea of *contingency*, or chance; luck of the draw. He believed that if we could "rerun the tape of life" from any point in evolutionary history, things would turn out completely differently. If we could start over 543 million years ago when the Cambrian explosion was getting underway and let evolution play out again, we would not end up with humans, or mammals, or even insects. The philosophical consequences of such a view are enormous: it strips human existence of any possible meaning or significance, reducing it to a very long string of lucky breaks that could never be repeated.

Conway Morris countered this view in his book *The Crucible of Creation* (1998), criticizing Gould almost savagely. This too was the story of the Burgess fossils, but Conway Morris reversed his earlier conclusion that they represented an explosion of new kinds of life. After further study of the many strange animals, Conway Morris and others realized that they had misinterpreted some of the fossils, and that they could be grouped into a smaller number of phyla, meaning that the Cambrian episode was *not* such an unusual explosion-like event; and newer evidence showed that there have been many new phyla since the Cambrian; that is, the total number of phyla did not decrease after the Cambrian episode, but has actually increased.

The sharpest difference between Gould and Conway Morris concerned the idea of rerunning the tape of life. Conway Morris believed that things would come out very similarly every time, that evolution in different lineages tends to converge at similar solutions to the problems of survival. This is known as *convergent evolution*.

There are many examples of evolutionary convergence. Dolphins and sharks arrived independently at very similar body shapes, beginning from completely different lineages. Dolphins evolved quite recently from dog-like mammals and sharks are among the most ancient fishes, yet both arrived at the same solution to the problem of drag while moving through water (as did every fish). Placental mammals and marsupial mammals both produced a large, saber-toothed carnivore on separate continents. Perhaps the most striking example of convergent evolution is the appearance of the "camera" eye, with a lens and screen (retina) like ours. The lineage leading to squids accomplished this, and so did jellyfish, snails, spiders, some marine worms, and all the vertebrates. The evolution of these diversely separate lineages can be thought of as a different running of the tape, and for each run the end result was very similar. This end result, a sophisticated eye, was not the product of an unrepeatable string of lucky breaks as in Gould's contingent evolution. It was produced by natural selection favoring something that gave significant survival advantages, and it's clear that the same innovation, if it's a good idea, can be arrived at many times from very different starting conditions.

Convergent evolution is now widely accepted among biologists, though the idea is not new—even Darwin noticed it. Yet we cannot dismiss the role of contingency, or chance. After all, it was the chance collision of a meteor with the Earth 65 million years ago that eliminated the dinosaurs and opened the door for the mammals, and ultimately humanity. We must conclude that convergence and contingency are both important aspects of biological evolution.

Like Gould, who took the extreme position of radical contingency, Conway Morris also occupies an extreme at the other end of the spectrum. In his 2003 book, *Life's Solution—Inevitable Humans in a Lonely Universe*, he says that biological convergence provides clear and decisive evidence that evolution is limited in its possible trajectories and, in fact, very few things may actually be possible. He goes so far as to say that evolution follows inevitable and preordained trajectories, that it is goal-directed; and the strong implication is that the goal is humanity. Evolution, one way or another, will eventually converge on conscious, self-aware beings. Replay the tape any number of times, starting anywhere at all in the long history of life on Earth, and the result would still be humanity.

Of course this suggests a plan, if not a Creator, behind biological evolution, and puts humanity in a place of central importance: humanity is the pinnacle and the purpose of evolution. The goal of life on Earth was to create us, and natural selection was the mechanism that brought it about. Many scientists say that Conway Morris has gone beyond science in this view, and ventured into religious belief. Indeed, neither Conway Morris nor anyone else has been able to put forward a rigorous scientific argument for the inevitability of evolution, that if we could rerun the tape we would end up at largely the same place. Yet no one has made a scientific argument that refutes this view, so it remains possible.

We are left with no answer to the age-old question about the status of humanity. Are we the result of a long string of lucky breaks that could never play out in the same way again, as Gould believed, so that there is no significance or meaning to human existence? Or are we the inevitable, pre-ordained result of biological evolution, so that we are the most important thing on Earth, if not the universe? The reader must decide.

Chapter 7
Living on Land

Life on Land: The Age of the Reptiles

Geocosmic Time: Day Three, 9:30 pm to 11:40 pm

Our story of life so far has taken place entirely in water. For more than 3 billion years—from about 5:00 in the morning until about 9:30 in the evening of day three in the life of the universe—life existed only underwater. Finally, about 475 million years ago, after the Cambrian explosion, life made the move to land.

Earth's landmasses had been in a state of flux since the first continents emerged from the oceans about 3 billion years ago. Because continents are giant floating plates, they are constantly moving. It is not known what the positions of the continents were billions of years ago, but the last half-billion years is fairly well understood (Figure 7-1 next page). When life started moving onto land, there was probably just one large continent, Pangaea, that later split apart.

The move out of water and onto land was a huge step. The environment on land was comparatively harsh—hot and dry, and bathed in ultraviolet radiation. Tough outer skins evolved for protection. Also problematic was the absence of the buoyancy that water provided. A new and more robust structural plan had to be developed for both plants and animals to support their own weight on land; land animals also had to develop lungs for breathing air, instead of using gills to extract oxygen from water.

It was plants, originating from green algae, that first ventured onto land about 475 million years ago. These first land plants were Bryophytes, much like the mosses of today. Evolving from these, around 400 million years ago, were fern-like plants with stems but no seeds; at about 380 million years ago plants with seeds evolved, including cone-bearing plants; then very recently, about 130 million years ago, the familiar flowering plants emerged. When plants began to live on land, the atmosphere changed significantly because of the oxygen they produced. By about 350 million years ago when giant ferns covered the land, oxygen levels may have peaked at about 35%, compared to today's 20%. This allowed plants and animals to grow to immense sizes.

The first animals apparently made the transition to land about 415 million years ago. They were invertebrate arthropods that probably split off from the ocean-dwelling trilobites, much like the "rolly-pollies" (pill bugs) that we find under rocks today. These first land-dwelling arthropods diversified into the wide variety of insects, spiders, centipedes, and mites that have remained highly successful since then.

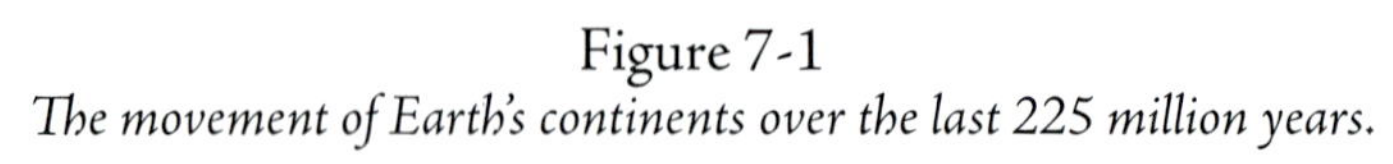

Figure 7-1

The movement of Earth's continents over the last 225 million years.

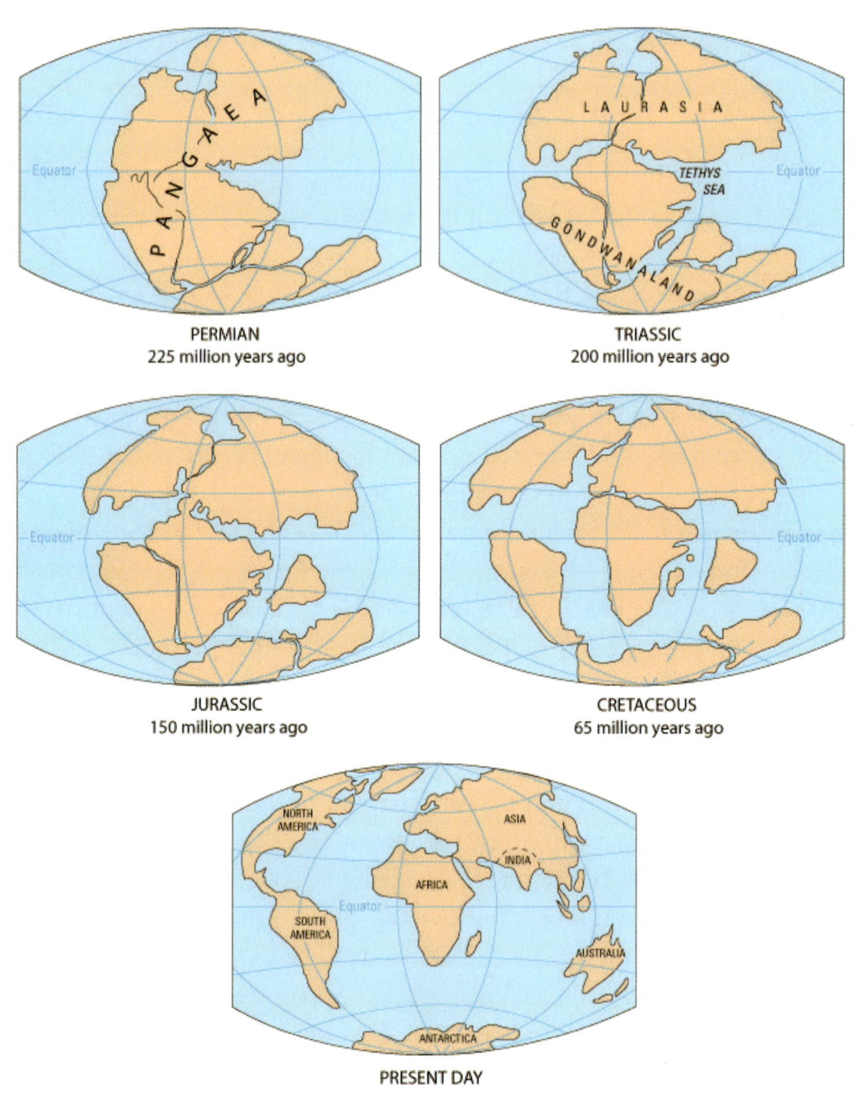

U.S. Geological Survey

The move to land came later for animals with backbones—the *vertebrates* or *chordates*. Emerging from the Cambrian explosion were worm-like animals with a *notochord*. This was a tube running the length of the body that eventually evolved into a backbone and spinal chord. These early chordates became the first fishes by about 510 mya, and by about 365 mya a certain lobe-finned fish apparently began to venture out of water and onto the mud flats of the land. Their fins developed into powerful limbs that allowed them to prop themselves up and slither around on land. These became the first amphibians, perhaps similar to the salamanders of today, spending time both on land and in the water.

Over the next 50 million years the amphibians gradually solved the problems of living in the open air. They developed lungs to breath the air and tougher skin to protect them from the dryness and the Sun's radiation. They also learned how to reproduce on land, which meant being able to mate and then lay eggs with protective shells. The animals that acquired these abilities became the reptiles, the first full-time land-dwelling vertebrates.

By about 320 mya the fossil record shows that the amphibians and reptiles had become a sort of mega-dynasty. They were the most complex and powerful form of life on Earth and would remain so until things changed abruptly 65 mya. The early reptiles eventually diversified into today's turtles, snakes, lizards, crocodiles, and alligators, as well as many extinct forms. Then, sometime around 230 mya, a new reptile appeared and some of these grew to enormous sizes. These were the dinosaurs and they would be the dominant form of life on Earth for the next 170 million years.

In the end, however, bigger was not better; the dinosaurs vanished after the K-T extinction event 65 million years ago. Yet one remnant of the dinosaurs still lives today. Fossils of the 150-million-year old archaeopteryx reveal a dinosaur with feathers that walked on two legs. The feathers probably originated as a way to regulate body temperature, but eventually archaeopteryx evolved into something with feathered wings that enabled flight. Aves, the birds we see everywhere today, are apparently the only surviving descendants of the dinosaurs.

The Rise of the Mammals

Geocosmic Time: Day Three, Beginning about 10:42 pm

At about the time the first dinosaurs appeared, there was another split from the reptilian line. These were the theropsids, or mammal-like reptiles. The theropsids would become a parallel evolutionary experiment in competition with the dinosaurs; and the dinosaurs, with their sheer size and power, easily won the competition. While the dinosaurs grew increasingly dominant, the theropsids evolved into small, fury, burrowing mammals that were hardly noticed by the dinosaurs. Yet in the end, the mammalian experiment was a far better plan: the mammals would ultimately inherit the Earth.

The first mammals appear in the fossil record about 248 million years ago, and mammal-like reptiles were probably evolving long before that. Early mammals are recognized by a jaw that was growing more powerful for chewing and teeth that were differentiating into incisors and molars; reptiles have a mouthful of nearly identical pointed teeth. One of the most important early innovations of the mammals was the development of lactation—the ability of a mother to produce milk for her young. This began the steady transition away from egg laying and toward live birth.

While the dinosaurs ruled, the inconspicuous mammals were quietly evolving in other ways. By about 180 mya the *monotremes* appeared in the southern continent of Gondwanaland (Figure 7-1, previous page). Monotremes lay eggs and produce milk, representing a transitional form between reptiles and later mammals. The only monotremes surviving today are the platypus and the spiny anteaters, found only in Australia and New Guinea. As the continents broke apart, the monotremes were stranded on these islands (University of California 2002).

Before Australia separated completely, the marsupials appeared on the continent of Gondwanaland. These were mammals with pouches, like the kangaroo, that did not lay eggs but gave live birth to their young. In marsupials, the gestation period is short and the newborn fetus, really an embryo, finishes its development in the mother's pouch where it is nourished by milk-bearing nipples. Unlike the monotremes, the marsupials spread to all the continents. Yet the only marsupial that survived outside of Australia was the opossum. In Australia, where kangaroos, koalas, and wallabies still roam, there may have been little competition for the marsupials, while on the other continents another type of mammal evolved and probably out-competed them. These were the placental mammals.

The placentals include all of the common hair-covered animals we think of today, such as horses, dogs, lions, elephants, humans, and even whales and dolphins. In all of these mammals, the developing embryo is nourished inside the mother through the placenta, which intermingles the blood supply of the mother and the fetus. The completed transition from egg laying to fully developed live birth conferred significant survival advantages – parents no longer had to bring food items back to an unguarded nest and could devote more time to nurturing fewer young. This probably contributed to one of the most important advances that mammals made over reptiles: a larger brain. Mammals developed sophisticated parenting behaviors to raise their young, in contrast to the reptiles that in some cases eat their newly hatched young.

In parallel with the larger brain, the mammalian jaw evolved an intricate 3-bone mechanism, the middle ear, which brought acutely sensitive hearing. The large brain could process the flood of auditory information into a 3-dimensional world of sound that reptiles never achieved. The sophisticated nervous system of mammals enhanced all of the senses and the brain could process much more sensory input. A further advance achieved by the mammals was the ability to maintain body temperature, in both cold and hot environments. They could live in a much wider range of environments than reptiles. Coats of fur were far better for maintaining body temperature than the outer skin of the reptiles, and a four-chambered heart could pump heated blood to every part of the body through an extensive circulatory system. The warm-blooded mammals could live in very cold environments, as long as they kept eating (they convert food to heat), while the cold-blooded reptiles would simply shut down and go to sleep when it got too cold.

Mammals had apparently acquired all of these advances by 100 mya, yet they remained almost powerless against the less intelligent dinosaurs. In a crucial stroke of luck for the mammals, the dinosaurs were suddenly eliminated by the impact of a meteor 65 million years ago, the so-called K-T (Cretaceous-Tertiary) extinction event. No one understands exactly why the mighty dinosaurs were completely wiped out while the meek little mammals survived. Perhaps the mammals burrowed underground for protection and maybe their diverse diet helped them. With the dinosaurs gone, the way was cleared for the mammals to flourish. Yet the fossil record shows that they remained small for another 10 million years. This slow fuse finally led to an explosion in diversity and size starting about 55 mya that has come to be called the *mammalian radiation*. Why the delay? Why did the mammals remain small for so long (200 million years)?

At least part of the answer was published in 2005 by a team of atmospheric chemists (Falkowski 2005) who discovered that while the dinosaurs lived, oxygen levels in the atmosphere were at only about 10% (compared to today's 20%), then rose to about 23% as the mammalian radiation was unfolding. The low oxygen level was fine for the dinosaurs, but larger mammals need three to six times as much oxygen as reptiles. When oxygen levels rose and the mammalian radiation finally began, the fossil record becomes full of large and exotic creatures like 4-foot tall guinea pigs, raccoons as large as bears, and 800-pound birds that could kill anything; other large mammals like saber-toothed tigers, mammoths, mastodons, and giant sloths appeared much later. Ultimately the experiment with being large was not successful, not for the dinosaurs or for the mammals.

The mammalian radiation was more about diversity than size, as thousands of new mammal species appeared. One of these was a squirrel-like mammal that was adapted for living in trees—it had hands and feet that could grasp branches and manipulate objects, and stereoscopic vision that required an even larger brain. These were the first *primates*. Early primates, called *prosimians*, have been found nearly world-wide, but as the continents broke apart and the climate cooled, they migrated to the tropics and became more monkey-like. By about 20 million years ago, primate evolution was booming on the continent of Africa, and in the final part of our story, Africa becomes the center of action, the incubator for every new species and every new development that led to modern humans. The evolutionary tree of the primates grew and diversified over tens of millions of years with new branches that included monkeys, gibbons, macaques, baboons, orangutans, and gorillas—the great apes.

Then, about 6 or 7 million years ago, at about 3 minutes before midnight on the third day in the life of the universe, part of the great ape lineage split into two: one became the chimpanzees and the other became the hominids—the upright-walking apes. The hominid line blossomed into at least twenty different species, but in the end all of these species went extinct—except one. In the next part of our story we will examine the intricate and mysterious saga of the hominid lineage that in the end produced one surviving species: *Homo sapiens*, also known as *humanity*.

Figure 7-2
The Evolution of Animals

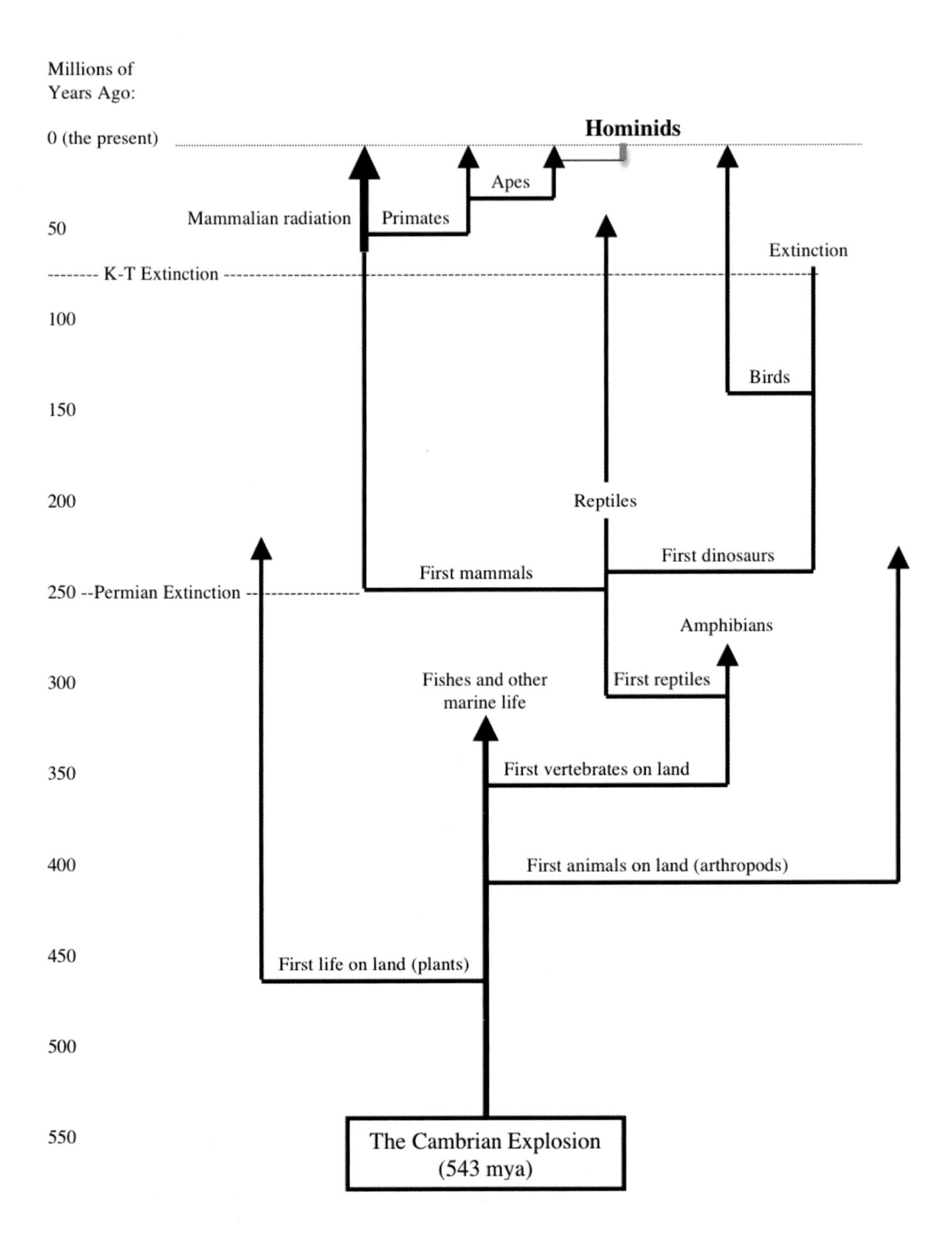

Exploring Deeper

The Brain

The evolutionary explosion of the last 500 million years produced a wide range of new innovations including fins, legs, wings, sophisticated sensory organs, circulatory systems, digestive systems, and skeletons. Probably the most important of all the innovations, however, was the brain and nervous system that began in early animals, and culminated with the human brain. It has been said that the human brain is the most complex phenomenon in the known universe.

From the beginning, the simplest forms of life had chemical communication mechanisms within cells. These grew in complexity as cells evolved, but not until about a billion years ago, with the advent of multicellular life, did the first simple nervous systems appear. In multicellular animals, cells began to differentiate into new types that could carry out specific functions. One type, called the *neuron*, could carry electrical impulses and took on the job of rapid communication within the organism. The first simple neural networks were made up of identical neurons connected together like wires running throughout the organism. This kind of primitive nervous system can be seen today in jellyfish. Their neural network mainly coordinates the opening and contracting of the skirt that propels it through the water.

Gradually, simple neural networks evolved into a central nervous system consisting of nerve networks tied to a central control center. Worms probably had the first central nervous system with a small brain-like control center connected to a cord of neurons running the length of the body. However, the tiny brain of a worm is barely necessary—worms can carry out most functions even when the brain is removed.

By about 410 million years ago, the first insects had developed a sophisticated brain that could coordinate the workings of a complicated body with many moving parts and internal systems. Although this brain was tiny, it enabled insects to master walking, crawling, swimming, hopping, burrowing, and flying. It could process input from many kinds of receptors, making them sensitive to odors, sounds, light patterns, texture, pressure, humidity, temperature, and the chemical composition of their surroundings. The insect brain brought many sophisticated behaviors. Leafcutter ants harvest leaves, then bring them into their nest where they cultivate gardens of edible fungus; honeybees live in highly structured social communities with strict division of labor, and can communicate information about food sources through a kind of dance.

With the emergence of the first vertebrates—fishes, amphibians, and reptiles—the brain became larger and more complex. The spinal cord was now much thicker and was protected inside the hard vertebrae of the backbone. It served as a two-way conduit of information, with sensory information traveling toward the brain and motor commands traveling out to the muscles. The early reptilian brain was a group of swellings at the top of the spinal cord devoted to survival matters. It managed involuntary actions like breathing and the beating of the heart, and some voluntary functions including the fight-or-flight response, sexual drives, and territorial instincts. An emotional center, creating fear and rage, also seems to have evolved in the reptilian brain.

The evolution of the brain in higher animals was hierarchical, both in form and function. The first mammals maintained the reptilian part of the brain that was still essential for survival, but added new structures over this older core. The most important new feature was a cortex where higher levels of thinking began to take place. Mammals took on a whole new set of behaviors. The parenting instinct was non-existent in reptiles, but for mammals, the young arrived helpless and vulnerable, making parental affection essential for survival. The cortex of the mammalian brain also got larger as the five senses became more refined, and at some point the brain could produce a 3-dimensional representation of the outside world by merging together the vast streams of data from the senses. Predatory mammals had a larger cortex than non-predatory mammals, allowing them to strategize and outsmart their prey. The functioning of the more advanced brain was also hierarchical, with an interplay of the older reptilian instincts and the more refined and reasoned mammalian perspective. This is still evident in humans today as anger and rage from the reptilian brain conflict with the reasoned thought generated by the mammalian cortex. In the most desperate of survival situations, the reptilian fight-or-flight response will still override reason.

By the time primates evolved about 50 million years ago, the cortex had grown so large that it completely engulfed the older reptilian core. The primate brain enabled sophisticated vision and hearing, both with depth perception and an acute directional sense of smell. The abilities to think and feel and solve simple problems were also emerging in the primates. However, as we will see in the next chapters, the ultimate explosion in brain growth and function was yet to come. The brain would quadruple in size as apes evolved into modern humans, and its internal wiring somehow organized to a level far beyond the biggest and best super-computers of today.

The cortex of the modern human brain is made up of some 100 billion neurons, with each one having about 10,000 connections with other neurons; and now researchers have found that within these neurons is a much smaller and denser system of microtubules that is even more complex. Somehow these interconnected microtubules and neurons create our ability to think, remember, make decisions, visualize and imagine, see the world, feel pleasure and pain, learn, and experience self-aware consciousness. Neuroscientists are barely beginning to understand these processes. The theme of brain development will continue to be a central part of our story as hominids evolve into fully modern humans.

The Upside of Extinction

Geocosmic Time: Day Three, 9:40 pm to 12:00 midnight (the present)

The evolution of life over the last 500 million years is truly astounding, as microscopic organisms transformed into magnificent plants and animals. One of the most important mechanisms that shaped evolution was not something that fostered life, but something that destroyed life: large-scale extinction events. We now know of five major extinction events since the Cambrian explosion, and many smaller ones. At many different times, life on Earth was nearly eliminated, but those few life forms that did survive each time evolved into new and more complex forms. The resilience of life is extraordinary.

The first of the five great extinction events, the *Ordovician extinction*, occurred about 444 million years ago. By some estimates it eliminated about 85% of all species (almost all of these were living in the ocean) (Sheehan 2001), and cleared the way for larger fish and aquatic reptiles. Some of the animals that survived this event later colonized the land, and so it seems that the Ordovician extinction catalyzed the move to land.

The second great event was the Late Devonian extinction, some 374 to 365 million years ago, wiping out perhaps 75% of the marine species, but apparently no land-living species (plants and insects at this time). It has been theorized that a global ice age was triggered by the emergence of vascular plants (plants with stems that circulate resources) on land (Algeo et al. 1995). With the development of vascular stems, plants grew in size from 30 centimeters to as much as 30 meters (think trees). This 100-fold increase in size was accompanied by a similar increase in root structure, which may have enhanced weathering processes on land; and increased weathering tends to sequester carbon as limestone in the ocean, which in turn leads to a drop in CO_2 levels in the atmosphere and thus a drop in global temperature (see The Global Thermostat and Plate Tectonics, page 94-96). Yet whatever caused this extinction event, for some reason it only affected marine life, and not life on land. In the rebound from this extinction event, atmospheric oxygen levels reached 35%, supporting huge plants and animals on land.

The greatest of all extinction events was the Permian extinction, 251 million years ago, in which some 95% of all marine species and 70% of land species were eliminated (The Great Dying 2002). This is the only one of the five that caused a mass extinction of insects. While some scientists think the event happened gradually over millions of years, others suspect that a sudden event like a meteor impact was the cause. No one yet knows. Before life could fully recover from this catastrophe, another extinction event struck 50 million years later, or about 200 million years ago. This was the Triassic-Jurassic extinction that eliminated about 50% of the species on Earth, and hit large animals especially hard. Again, the causes are unknown. These two extinction events, happening in relatively quick succession, opened the door for dinosaurs to become dominant. The large-scale pattern, then, seems to be extinction followed by mega-dynasty. Paradoxically, extinction events stimulated evolutionary progress.

The fifth and most recent extinction event was the Cretaceous-Tertiary (K-T) extinction that occurred 65 million years ago. This was one of the smaller extinction events, eliminating about 70% of Earth's species, but it was particularly significant for us as mammals. The K-T event brought the end of the dinosaurs* and opened the way for mammals to finally begin their mega-dynasty. This event is also unique because it is the only one of the five great extinction events for which we know the cause. The first four are still mysteries, though numerous causes have been suggested, including changing sea levels, meteor or comet impacts, global warming, global cooling, oxygen

* The avian dinosaurs survived the K-T event, for unknown reasons, to become today's birds (Aves).

depletion in the seas, epidemics of disease, massive volcanic activity, solar flares, super novae, and reversals in the Earth's magnetic field, to name a few. We do not know the duration of these events, although fossil evidence suggests that they probably lasted for millions of years.

However, the K-T event that wiped out the dinosaurs is now widely believed to be primarily the result of a meteor impact near the Yucatan Peninsula. This idea was first published in 1980 by Luis and Walter Alvarez and associates from the University of California at Berkeley (1980). They found a thin sheet of iridium between rock layers in Italy that dated to the time of the dinosaur extinction. Iridium is very rare naturally in Earth's crust and the Alvarezes proposed that the iridium they found must have arrived on a large extraterrestrial body because iridium is found in much higher levels in some kinds of meteors. The impact of a large meteor would have spread the iridium all over the Earth as part of a great cloud of dust. The iridium layer has since been found worldwide in 65 million-year-old deposits, lending support for the Alvarez K-T extinction theory. However many other explanations for the extinction of the dinosaurs have been put forward since 1980, including: a super nova explosion, extensive volcanism, pollen from newly evolved flowering plants, the dinosaurs got too big, mammals out-competed them, mammals ate their eggs, and that they slowly faded into extinction. Still, none of these have found any real support.

In 1978 an impact site, the Chicxulub crater, was discovered by petroleum geologists working in the Yucatan Peninsula of Mexico, but not until the 1990s was the crater studied more closely and dated to about 65 million years ago, strongly linking it with the K-T event. Yet some scientists were still not convinced, until finally in 2010 an international panel of 41 scientists endorsed the theory that the Chicxulub impact was the culprit (Schulte et al. 2010). After 30 years of debate, the Alvarez extinction theory was finally accepted.

The characteristics of the Chicxulub crater indicate that the impact object was 10 to 15 kilometers in diameter and traveling at about 10,000 miles per hour when it slammed into Earth, unleashing the energy of a billion or so Hiroshima-sized atomic bombs. The immense heat ignited sulfur-bearing rocks in the seabed and hurled vast quantities of sulfuric acid into the atmosphere, which fell as toxic acid rain. Sunlight was blocked out all over the Earth by the smoke and other particulates that were thrown into the atmosphere. This made photosynthesis almost impossible, which eliminated many green plants; thus the dinosaurs that fed on green plants could not survive.

The K-T event was very selective: 90% of the species that perished were land animals, while only 10% were marine species. Land plants were hit very hard, so virtually all land-dwelling plant eaters died, while scavengers that fed on a variety of things like insects, worms, snails, and larvae survived. Largely spared were fishes, bird-like dinosaurs, non-dinosaurian reptiles, and the tiny, furry mammals that had been living unspectacularly on land throughout the reign of the dinosaurs. With the dinosaurs gone, the door opened for the mammals to become the most dominant form of life on Earth. Once again, extinction opened the door for new, more complex forms of life.

The mammals have now ruled the Earth for the last 55 million years, but the evidence is growing that the Age of Mammals is ending. Abundant fossil skeletons show that many types of large mammals roamed North America and the other continents thousands of years ago: mastodons and mammoths (close relatives of the elephant), sloths, armadillos, jaguars and saber-toothed tigers, many species of horses, giant bears and beavers, peccaries, and musk ox, to name a few. Virtually all of these are gone.

Biologists in the 1970s began to notice drastic declines in many animal populations all over Earth as human overpopulation was becoming apparent for the first time. Analysis of the fossil record shows that the average rate of extinction for mammals over their entire history has been about one species every 400 years. So it was very alarming when mammologists from the American Museum of Natural History (MacPhee 1999) found that about 100 species of mammals had become extinct over the last 500 years—about 80 times the average rate; and some estimates place the current rate of extinction of mammals at 100 to 1000 times the average (Lawton and May 1995). If we consider all types of life, we may be losing about 140,000 species per year (Brooks et al. 1995).

Most scientists now agree that we are in the midst of a sixth great extinction event that has come to be called the *Holocene extinction*, named after the geological time period that began about 12,000 years ago. However this one is different from all the others. First of all, it is being caused by a single species of life—humans—while all of the preceding events were geological in nature. Secondly, it is happening faster than any of the others. While the first five events probably took thousands or even millions of years to unfold (even the K-T meteor impact took thousands of years to play out), we may eliminate 50% of all species on the Earth by the end of this century. It remains to be seen whether or not humans can take action quickly enough to slow down or stop the Holocene extinction, or whether this extinction event, like the others before, will clear the way for something new.

Part Three Summary

Geocosmic Time: Day Three, 11:00 am to 11:57 pm

For its first billion years or so, life on Earth existed only as microscopic, prokaryotic organisms—primitive bacteria that used carbon dioxide and died in the presence of oxygen. In these early days, Earth had a CO_2 atmosphere. Then, as some prokaryotes started using photosynthesis as a source of energy, CO_2 in the atmosphere was depleted and replaced by toxic oxygen. Life was being poisoned by oxygen; and to make matters worse, the drop in atmospheric CO_2 triggered the worst deep-freeze of all time: snowball Earth. Eventually a thick layer of ice grew to cover almost the entire planet. At this time, about 2.5 billion years ago, things were not looking good for life on Earth. Yet somehow, some forms of life developed the ability to use oxygen in a new energy process we call cellular respiration.

In the face of a toxic atmosphere and a frozen Earth, primitive life did a remarkable thing: larger organisms merged with smaller ones in a collaborative arrangement that benefitted both. Bacteria that had mastered oxygen breathing were assimilated to become the mitochondria that animals use for a power source, while small photosynthesizing organisms were engulfed to became the chloroplasts of green plants. This collaborative arrangement was the beginning of eukaryotic life, and it opened the door that led to all of the large, complex life forms we know today. The eukaryotes not only solved the oxygen crisis, but also attained a much higher level of internal organization and sophistication; perhaps most importantly, they developed sexual reproduction, significantly accelerating the pace of evolution. Even so, the eukaryotes would remain as one-celled protists for another billion years or so while reaching the complexity level, in one cell, of a Boeing 747.

Sometime about 1.5 billion years ago, life once again used the strategy of collaboration. This time eukaryotic cells began to join together, with different cells taking on specialized functions. Multicellular life began. Although it remained microscopic for another billion or so years, there was great experimentation and innovation steadily going on, bringing about new structures like simple nervous systems and the clear distinction between plants and animals. Then, about 543 million years ago, during the Cambrian period, as atmospheric oxygen levels were peaking, these tiny creatures exploded in size and diversity; suddenly the fossil record is full of large, complex plants and animals. The Cambrian explosion was the beginning of modern life, even though life still lived only in the oceans.

The next great step for life was to make the move out of water and onto land. Plants were the first to make this transition about 475 million years ago, and 60 million years later the first simple animals walked on land. Now the pace of evolution was frenetic compared to the early days, and spectacular new types of plants and animals emerged: flowering plants, flying insects, reptiles, fishes, dinosaurs, birds, and finally mammals.

Throughout this time there were five massive extinction events, each one eliminating the majority of species on the planet; yet each time, life came back, and it came back at even higher levels of complexity and sophistication. The most recent extinction event, the K-T event, happened about 65 million years ago, bringing the end of the dinosaurs and clearing the way for mammals to become dominant.

The mammals evolved into hundreds of different species that included enormous beasts like mastodons, saber toothed tigers, giant bears, and a clever newcomer, the primate. With an ever-growing brain, the primate lineage flowered into the great apes, and from the great apes a momentous split took place about 7 million years ago. From a common ape-ancestor came two lineages: one eventually became the modern day chimpanzees and the other—the hominids—culminated with modern humans. This new kind of mammal, the hominid, was neither large nor ferocious, but walked upright, had a large brain, and could use its grasping hands in totally new ways. The next part of our story lies in the field of paleoanthropology and follows the fragmented and still-mysterious journey from ancestral apes to modern humanity.

Part Four

The Age of the Brain

PART FOUR

THE AGE OF THE BRAIN
Becoming Human

When workers at a quarry in the Neander Valley of Germany blasted into a limestone cave in 1856, they stumbled onto the skeletal remains of a human-like creature. Local experts examined the discovery and found that it had a large skull, a thick brow with a sloping forehead, and heavy bones; it must have been large and strong. However it was not the first creature like this to be found—similar remains had been discovered in Belgium in 1829, and in Gibraltar in 1848, but this discovery became its namesake. The Neanderthal Man, later given the scientific name *Homo neanderthalensis,* was almost immediately controversial. Was this a thick-set modern human who had died in the last few hundred years, or was it an ancient ancestor who was not yet fully human? Did Neanderthals turn into modern humans or were they a parallel species that went extinct? And what was the relationship on the ground between humans and Neanderthals, who shared Eurasia for nearly 20,000 years?

The discovery of Neanderthal Man was the beginning of the science of paleoanthropology, one of the most contentious and colorful of all the sciences. The study of Neanderthals fills just one specialized area of paleoanthropology; more than this, paleoanthropologists try to piece together the whole story of how modern human beings emerged over the last few million years. Although this story is becoming more complete all the time, there are still many unanswered questions and longstanding mysteries about the origins of humanity.

But a warning ... Most of the science we have explored so far in this book is widely accepted and largely uncontroversial. From here on, however, things get much less certain, and controversy is everywhere. There are competing theories and different schools of thought about the events and players of the last 7 million years. Here we will attempt to present the most widely accepted views, while acknowledging alternative theories.

Time Context for Part Four

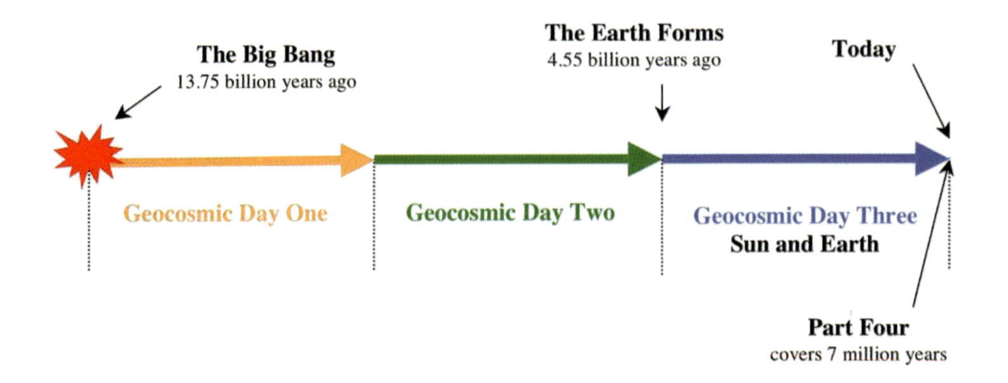

Main Events of Part Four: The Age of the Brain
All dates are approximate

Event	Years Ago	Geocosmic Time (all are Day 3)
Hominid lineage begins (Africa)	6-7million	about 11:58 pm
First *Australopithecus* (Africa)	4 million	11:58:45 pm
First *Homo* (Africa)	2.5 million	11:59:15 pm
First *Homo ergaster* (Africa)	2.0 million	
First *Homo erectus* (Eurasia)	1.9 million	
First *Homo heidelbergensis* (Africa)	1.2 million (?)	
Homo heidelbergensis migrates to Europe	1.1 million (?)	
First Neanderthals (Europe/Eurasia)	370,000	11:59.53 pm
First *Homo sapiens* (Africa)	200,000	11:59:56 pm

Chapter Eight
The Apes Who Learned to Walk

The Missing Link

At the time of the 1856 Neanderthal discovery there was no scientific explanation for human origins—this was still three years before Darwin published *The Origin of Species* in which he proposed that humans had descended from apes. The most widespread belief of the time in western culture was that God made this world, and all the living things, in a fairly sudden act of creation. Some interpretations of the Bible gave an age of about 6,000 years for Earth and its inhabitants, and this could not really be disputed in the 1850s because scientific dating methods were almost nonexistent. However, in the twentieth century, powerful new dating techniques were developed (see Appendix III), giving a much different figure for the age of the Earth and the age of the Neanderthals. Hundreds more Neanderthals have now been found, along with many of their artifacts, showing that they lived all over Europe, Central Asia, and the Middle East from perhaps 400,000 years ago until about 30,000 years ago, when they disappeared completely. How they interacted with early modern humans and why they disappeared is still uncertain. Later in this chapter we will explore these questions.

When Darwin first proposed the idea that humans descended from apes, it generated great interest in finding a "missing link" between apes and humans. In 1888 Eugene Dubois, a Dutch medical doctor, gave up a comfortable career in Europe to search for ape-human fossils in Indonesia. He had no idea where to look, but almost everyone at the time believed that humans originated in tropical Asia. He took his team of fifty men to the island of Sumatra because it was a Dutch colony. After several years of searching, he found thousands of animal fossils but no missing link; so he moved his search to the nearby island of Java. There, within a year, he found a skullcap, part of a femur, and several teeth that belonged to an individual that has come to be called the "Java Man." It had a lighter build than a Neanderthal, and a skull volume that was larger than an ape but smaller than a human or Neanderthal, so Dubois was certain that he had found the missing link. When he returned triumphantly to Europe with his find, however, it was dismissed as nothing more than a recent ape. It would be forty years before the mainstream scientific community realized that Dubois' Java Man was not an ape but an upright-walking hominid that lived hundreds of thousands of years ago. Was it the missing link?

Dubois was the first person to search deliberately for fossils of human ancestors, making him the first true paleoanthropologist. Since then, paleoanthropologists have searched all over the world for more bones, teeth, and tools that give clues about human origins. After more than 100 years, they have found and catalogued thousands of fossil specimens, mostly from East Africa and dating from as far back as 6 to 7 million years ago. However a single missing link has never been found, nor does anyone today expect to find it. Major discoveries over the last twenty years have solidified the picture of a continual transition from ancestral apes about seven million years ago to anatomically modern humans about 200,000 years ago, with many side branches that led to extinction.

The Hominid Puzzle

We know from many lines of evidence that the story of human evolution begins about six or seven million years ago when part of the lineage we call the *great apes* split into two separate lines (see Figure 8-1). One line became the modern chimpanzees (and their close relatives the bonobos) and the other line, the *hominids*, the upright walking apes, eventually led to modern humans (*hominids* are also called more formally the *hominins*).

Figure 8-1
The Evolutionary Tree of the Great Apes

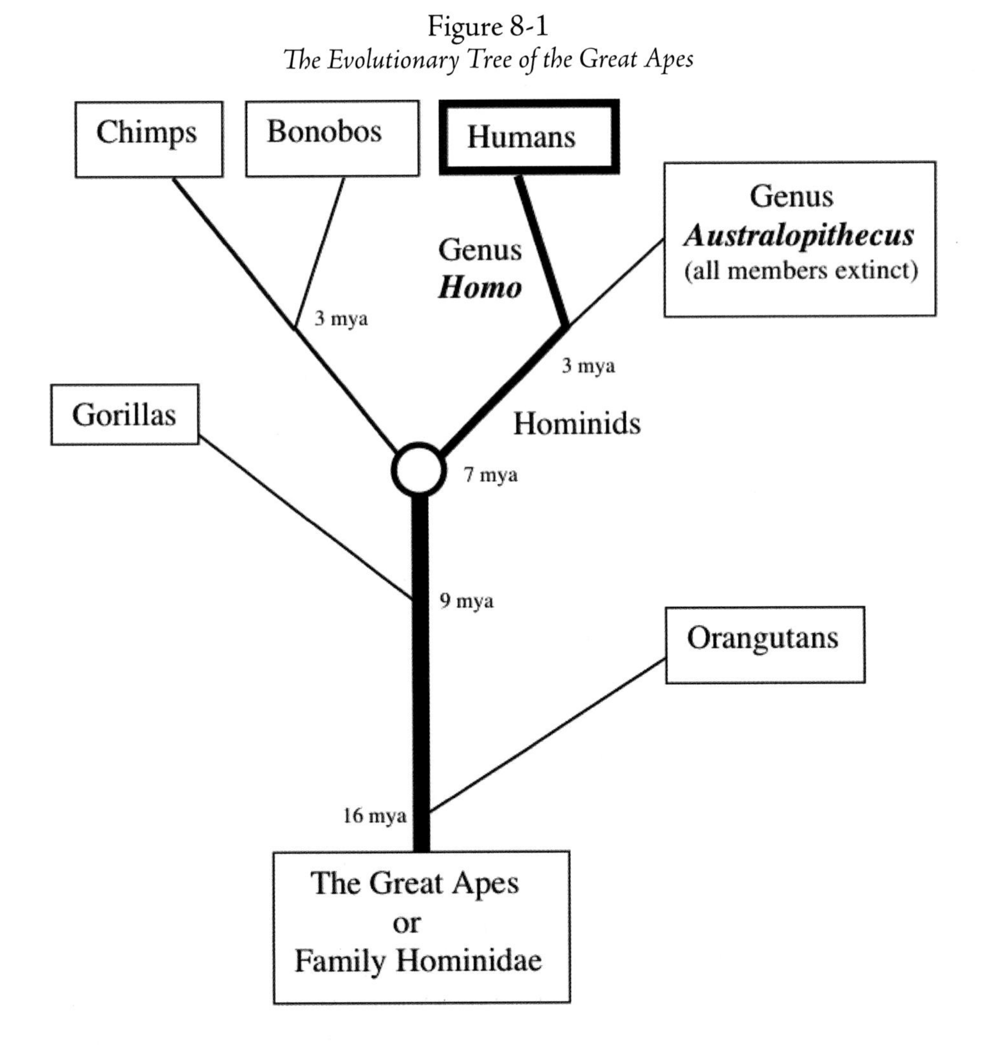

The story of human evolution is the story of this hominid lineage that split from the other Great Apes, and then split many more times into ten or twenty different species (depending on which paleoanthropologist you talk to); but eventually every one of these hominid species went extinct, except one: the only surviving hominid is *Homo sapiens*.

Most of our knowledge about how the hominid line evolved from apes to modern humans over the last 7 million years comes from the fossilized remains that paleoanthropologists carefully dig up, painstakingly clean off, and then protect in museums. Usually all that survives after millions of years are teeth and skull fragments or, in the best specimens, larger pieces like femurs and pelvic bones. Beginning about 2.5 million years ago, hominids began making stone tools that have survived almost perfectly until today; these provide many clues about intelligence and behavior.

Today there are over 5000 known fossil specimens of hominid individuals. This may seem like a large number but most of these are just a few fragments of bone and teeth, and 5000 individuals is a very small fraction of the many *billions* of hominids that once lived. For paleoanthropologists, putting together the story of human evolution is like trying to surmise the picture on a gigantic jigsaw puzzle by looking at just a handful of the pieces.

Scientists can learn a great deal from teeth and bone fragments. The teeth yield information about the diet and age of the individual, the skull reflects brain size and development, and the hip structure provides clues about how the creature moved, whether on all fours or upright. The Java Man that Dubois discovered in 1891 consisted of just a few fragments, yet that was enough to show it was not a modern human, or an ape, but something in between. Further study of these fragments has shown the Java Man lived about 700,000 years ago (there is still considerable uncertainty in this date) and was much more human-like than ape-like. We now call this hominid *Homo erectus*.

After Java Man was discovered in the 1890s the next big finds were in China during the 1920s. Johan Gunnar Andersson, a Swedish archeologist, heard about a place called Dragon Bone Hill near Peking (modern day Beijing), and when he was taken there, he found bone fragments and teeth scattered everywhere. The locals had been collecting them and grinding them into a powder for medicinal purposes. Andersson did some excavating and collecting, and took a few of the specimens to Davidson Black, a Canadian scientist working in China. Black immediately recognized their importance and launched a large-scale investigation of the site. Between 1929 and 1937, teams of archeologists found the remains of about 40 hominids, including six nearly complete skulls. This new find came to be called "Peking Man." In 1937 the Japanese occupied the area and took most of the specimens. They mysteriously disappeared and have still never been found. Fortunately, plaster casts had been made of the best pieces. When the war was over, digging resumed, and many more bone fragments belonging to Peking Man have since been found. Peking Man, like Java Man, is now classified as *Homo erectus*, and has been dated to about 750,000 years old (Stone 2009).

Also during the 1920s, an Australian anthropologist, Raymond Dart, unearthed a small hominid specimen in Taung, South Africa. The *Taung Child*, as it came to be called, was more ape-like than any of the other hominids that had been discovered at the time, and he named it *Australopithecus africanus*. Dart believed that his find was much older than either Peking Man or Java Man, and if this was true, it meant that humans must have originated in Africa rather than Asia. Others, however, dismissed his discovery as the remains of a gorilla because they believed so strongly that humans originated in Asia, not Africa. For the next 30 years most paleoanthropologists focused their search for human ancestors on sites in China and Southeast Asia. Not until 1959 did the work of Louis and Mary Leakey show that Dart had been correct in believing that hominids originated in Africa.

Louis and Mary Leakey were undoubtedly the most famous paleoanthropologists from the twentieth century. Starting in the 1920s Louis Leakey began to explore East Africa in search of the first human ancestors. He met Mary in 1933—she was an illustrator for his book *Adam's Ancestors*—and soon they began to work together in Olduvai Gorge, Tanzania. They dug at Olduvai for twenty-five years, finding many stone tools but no hominid fossils to support the theory of African origin. Finally in 1959 Mary found the first of many hominid fossils at Olduvai, while the more famous Louis was bedridden with fever. In the October 1961 *National Geographic* they introduced their new find to the world, naming it *Zinjanthropus boisei* (a name that would be changed to *Australopithecus boisei*, and later to *Paranthropus boisei*). They initially estimated that "Zinj" was 600,000 years old, but better dating techniques later showed that it was about 1.75 million years old. This made it the oldest hominid known at the time. It was clearly much older and much more ape-like than the Java and Peking specimens, and finally the theory of African origin had strong support.

Soon paleontologists from all over the world were flocking to East Africa in search of even older hominids. During the 1960s, the Leakeys also discovered *Homo habilis*, a new type of hominid dating to about 1.9 million years ago. However their work together was coming to an end. Louis and Mary Leakey parted ways in the mid-1960s after 30 years of working together, and Louis spent his remaining time in Europe fundraising and lecturing until his death in 1972. Mary continued working in Africa until she died in 1996, and their son Richard, his wife Meave, and their daughter Louise are still active researchers in Northern Kenya.

The 1970s brought a new generation of paleoanthropologists to Africa who continued to find fossils of hominid species that were even older. Donald Johanson and Tim White each made important discoveries in Africa and would become two of the best-known paleoanthropologists of the late twentieth century. In 1974 a team led by the young Johanson found a 40%-complete skeleton in the Afar region of Ethiopia. This ape-like creature that stood about three and a half-feet tall and weighed about 65 pounds came to be called "Lucy", after the Beatles song that was playing in camp near the time of discovery. Lucy, presumed to be female, was given the scientific name *Australopithecus afarensis,* and was dated to about 3.2 million years ago.

Was Lucy a hominid or an ape? That is, did she walk upright on the ground or move around by swinging from tree branches? Bipedalism is the defining characteristic of the hominid line and Lucy would be the oldest hominid yet discovered if it could be shown from her skeleton that she walked upright. Johanson and his team believed that Lucy's pelvic structure and the placement of her skull on her spine strongly suggested that she was bipedal. Then a few years later Mary Leakey and her young assistant Tim White confirmed that Lucy and her relatives must have been bipedal when they excavated a clear trail of hominid footprints at a site in Tanzania called Laetoli (Figure 8-2). The 70 footprints stretching over about 100 feet were made by two adults and a child walking across a layer of volcanic ash, and were preserved when the layer hardened. The layer was dated firmly to 3.5 million years old, leaving little doubt that the hominids of that time were bipedal.

Figure 8-2a
The 3.5 million year-old Laetoli Footprints made by Australopithecus afarensis.

From Photo Researchers/SPL

Figure 8-2b

Recent analysis of the Laetoli Footprints comparing them to a normal human walking (top) and an ape-like gait with bent knee and bent hip (BKBH)(Raichlen et al. 2010). The Laeotli Australopith walked very much like a human!

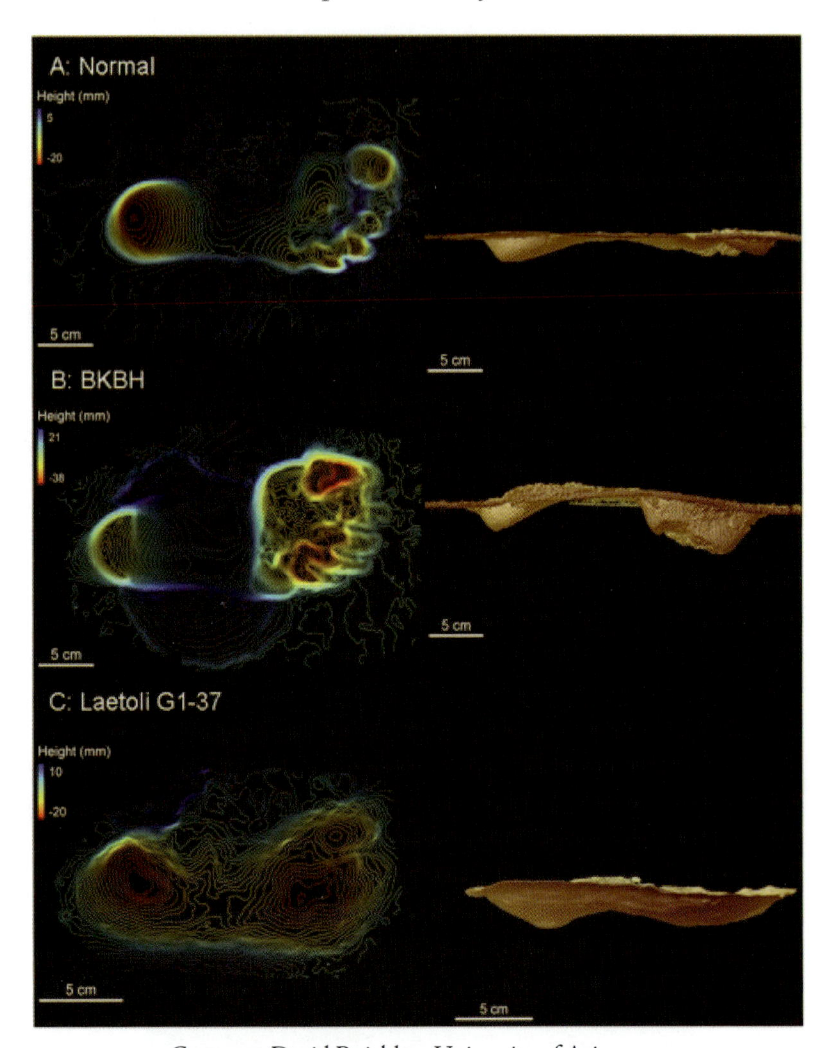

Courtesy: David Raichlen, University of Arizona

Neanderthal Man, Java Man, Peking Man, the Taung Child, Zinjanthropus, and Lucy—these were just the tip of the iceberg. By the late 1960s there were over one hundred named hominid species, creating a bewildering and chaotic picture of human evolution. Who were all these creatures and how were they related to each other, and to us? No one really thought there were a hundred distinct hominid species, but it was unclear how they all fit together.

Something had to be done to simplify and organize this mess, and the first step came when it was agreed that from about 4 million years ago to the present, there were broadly just two types of hominids, and these were given the genus names *Australo-*

pithecus and *Homo*. Within these two groups, however, there were still many species and unknown relationships. The organizing principle that created a more coherent picture was the evolutionary tree.

Science and Discovery

The Evolutionary Tree and Taxonomy

Hominid evolution and the evolution of all life on Earth begin to make sense when viewed as an evolutionary tree; the way in which scientists classify and name living things, called *taxonomy*, is based on the evolutionary tree of life. Today's system of taxonomy, using genus and species names, originated with the work of Carl Linnaeus in the mid-1700s, about a hundred years before Darwin solidified the concept of the evolutionary tree.

Biological taxonomy gives a two-part scientific name to every life form, past and present. The two parts are a genus name and a species name. Modern humans are in the genus *Homo*, and species *sapiens*. This two-part name is usually italicized when it is written, with the genus name capitalized and the species name in lower case. So our scientific name is *Homo sapiens*. The scientific name for the chimpanzee is *Pan troglodyte*; the domestic dog is *Canis familiaris*. However the genus and species are only the most specific levels of description. More completely, every life form belongs to a domain, a kingdom, a phylum, a class, an order, a family, a genus and a species. A complete classification of modern humans is shown in 8-3. You could say that the complete scientific name for you and me is *Eukarya Anamalia Chordata Mammalia Primatae Hominidae Homo sapiens*; but let's stick with *Homo sapiens*!

Taxonomy, however, is more than just a naming scheme—it represents a pathway on the evolutionary tree. Notice in Figure 8-3 that moving from the domain at the bottom to species at the top depicts evolutionary history. Going from the domain to the species corresponds to moving up the evolutionary tree, from the trunk to the ends of branches. A living species, occupying the tips of the branches on the evolutionary tree, is the culmination of a long evolutionary path. The complete taxonomic classification for humans—Eukarya, Animalia, Chordata, Mammalia, Primatae, Hominidae, Homo, sapiens—describes the unbroken evolutionary path that led to modern humans.

Until recently our understanding of the evolutionary tree was based only on *morphology*, or the physical appearance of organisms. For example, common vertebrates can be easily classified by looking at some very obvious physical characteristics: those that have fins and swim in water are fishes, those that have feathers and fly are birds, those that have hair and suckle their young are mammals, and so on. Today, in addition to morphology, we use DNA analysis to determine the evolutionary relationships between organisms, mapping out the evolutionary tree from the information stored in the genes of the DNA molecule. We now see the evolutionary tree as a map of how DNA has changed and moved through time, as life evolved. Darwin, standing on the shoulders of Linnaeus, made a giant leap when he recognized that life was organized in a many-branching tree, and that all living things share a common ancestry. This view (sometimes called the Theory of Evolution) is now strongly supported by the information found in DNA.

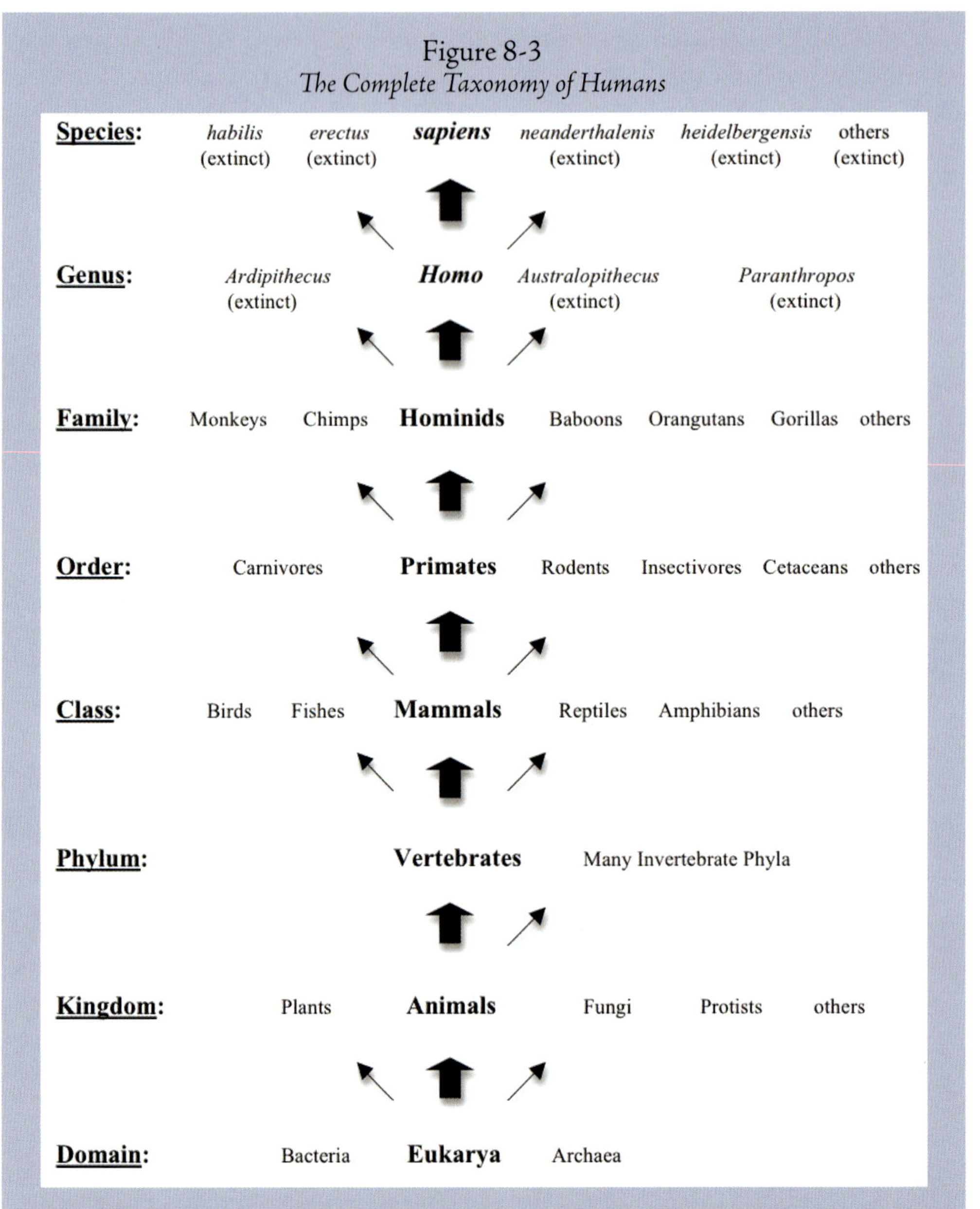

Figure 8-3
The Complete Taxonomy of Humans

Early evolutionary trees like Edward Hitchcock's from 1840 (Figure 8-4b) had only two branches at the base, plants and animals, because these were the only two forms of life that were recognized before about 1860. By 1866 Ernst Haeckel (Figure 8-4c) identified the three main branches of life as plants, animals, and protists, and this view stood for the next 100 years or so.

In Figure 8-4e, a modern evolutionary tree is shown, with the three domains of life that are currently recognized; it is called a *phylogenetic tree* and is based on both DNA analysis (gene mapping) and morphology. The single stem or trunk at the very bottom represents the "last common ancestor" of all living things, which we think lived about 3.8 billion years ago (Chapter 4). Notice that in this modern tree, plants and animals occupy an inconspicuous place at the upper right corner.

Each of the branches in this tree split many more times, into classes, orders, families, genera, and finally species. To depict the entire tree of all life on a single page would be very difficult!

Figure 8-4
The Evolutionary Tree of Life Through History

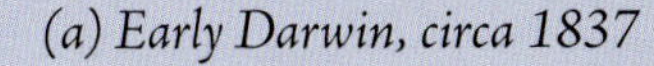

(a) Early Darwin, circa 1837

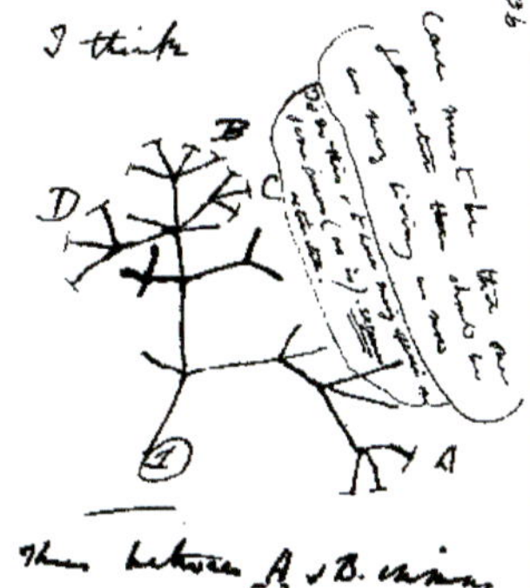

Wikimedia Commons

(b) Hitchcock, 1840

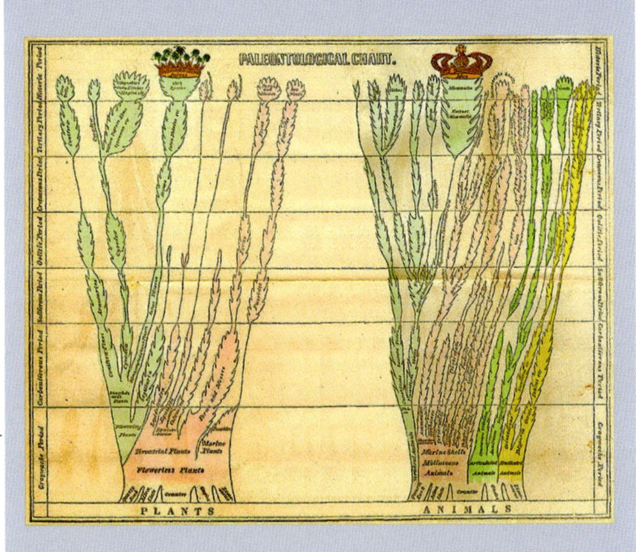

Wikimedia Commons

(c) Haeckel, 1866

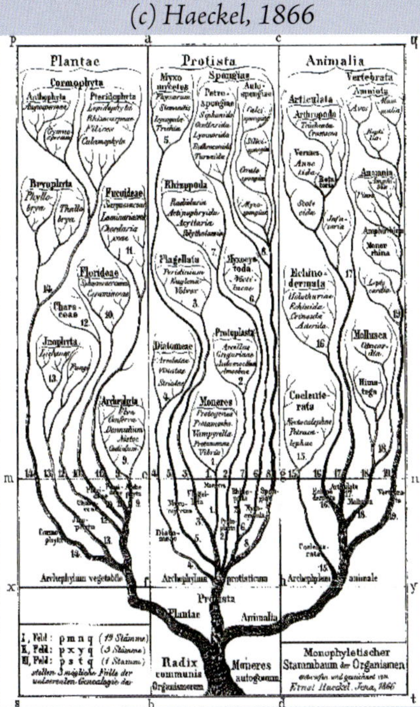

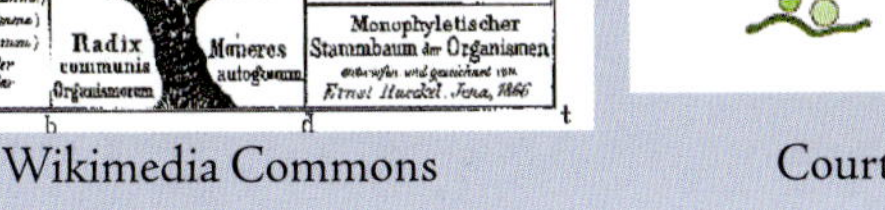

Wikimedia Commons

(d) Modern Pictorial

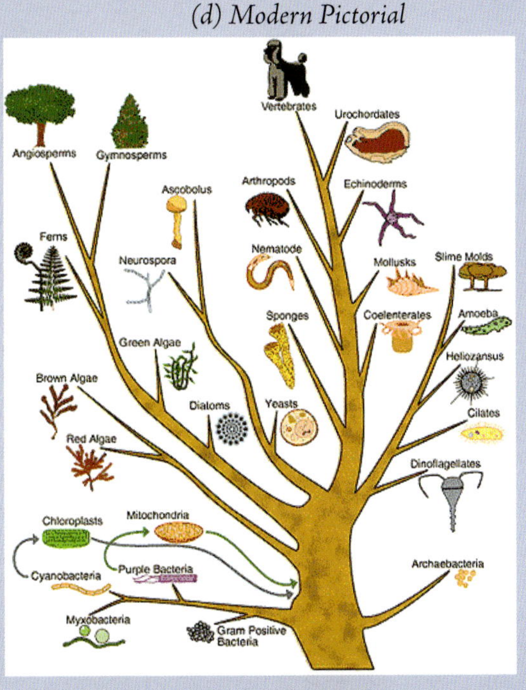

Courtesy: Consolidated Safety Services

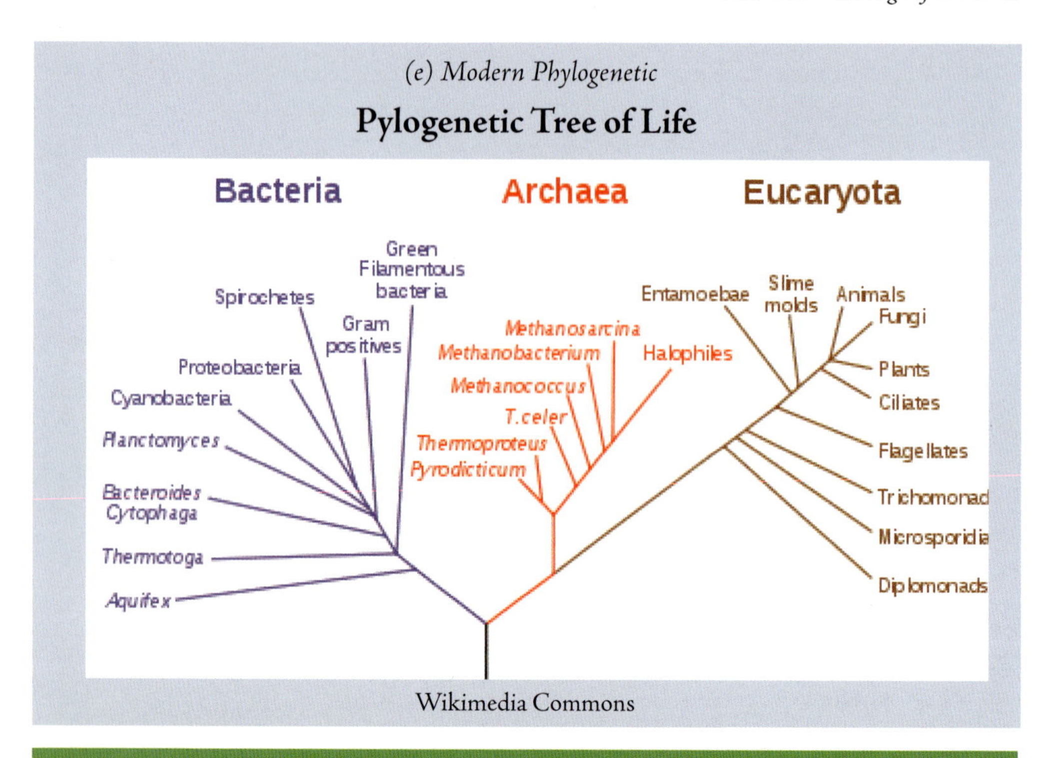

(e) Modern Phylogenetic

Pylogenetic Tree of Life

Wikimedia Commons

Exploring Deeper

Divergence, Speciation, and the Molecular Clock

As we move up the evolutionary tree, from the trunk to the tips of the branches, we are also moving through time. The history of life on Earth is recorded in the many branchings, with each split representing a new form of life. What causes a split on the evolutionary tree? That is, what causes new kinds of life to appear? This question is at the heart of how evolution works and is central to understanding hominid evolution. When a new lineage splits off from a branch of the evolutionary tree, it's called *divergence,* and the process that produces this split is generally referred to as *speciation,* or the making of a new species. Many lines of evidence suggest that a speciation event occurred about 7 million years ago when the great apes diverged into two lineages—the chimps and the hominids. What causes a divergence like this?

To understand speciation, we must first be clear about what a species is. It is, of course, the most specific grouping of life, in the way that we would probably think of roses as a species of plant. However this is not a workable definition in many cases. Although common sense might suggest that a horse and a donkey are so similar in their physical characteristics that they must be members of the same species, they are not; on the other hand, a Great Dane and a Miniature Poodle are so different looking that they would seem to be members of different species, yet they are the same species. More than similarities in appearance, the best criterion for membership in a species is *the ability to breed.* Biologists generally define a species as *a population that is able to interbreed with each other in nature and produce viable, fertile offspring.*[*]

[*] This definition of a species works fairly well for higher animals, but it does not work consistently for plants where separate species can sometimes be interbred, and it is even more problematic for simple forms of life like bacteria that share DNA quite freely.

Put more simply, an animal can mate and have healthy offspring only with members of its own species. The Great Dane and Miniature Poodle, if they find a way to mate, can actually produce a healthy fertile mutt, and therefore they are members of the same species. Yet the horse and donkey, although they can mate and produce a mule, are not members of the same species because the mule is infertile. The requirement for fertility in the offspring is not just an arbitrary rule—it's one of Nature's most important rules for living things: if you cannot pass along your genes, your lineage dies. So the horse and the donkey have no future lineage as interbreeding organisms. They are on different branches of the evolutionary tree, and the unfortunate mule is an evolutionary dead end because it cannot pass along its genes. In most cases, two organisms that are not closely related have different numbers of chromosomes in the cell nucleus (like humans with 23 pairs of chromosomes and goldfish with 40 pairs) and therefore cannot possibly interbreed. However, all humans on the Earth are members of the *same* species, because all can interbreed to produce viable fertile offspring; and we all have 23 chromosomes.

With our current knowledge of DNA, we can understand a species as a population that is so similar genetically that reproduction within the group is possible. Over many generations, however, the DNA within any population gradually changes—the slow random changes in DNA are called *genetic drift*. Eventually the DNA of two organisms whose ancestors were members of the same species may become different enough that they can no longer produce fertile offspring. If this happens, the two organisms would then be members of different species. A new species thus emerges (speciation) and a new lineage branches off of the evolutionary tree (divergence).

However, the process of speciation caused by genetic drift is very slow, probably taking millions of years to produce results. A much more powerful driver of speciation is geographic isolation, which biologists call *allopatric speciation*. To see how it works, let's imagine a population of gorillas becoming separated geographically—maybe a few members migrate to a distant place, maybe a cataclysmic event like a volcanic eruption drives the population apart, or maybe a few individuals get stranded on a log which floats across a body of water to a new land. The two separated populations will adapt to their new environments through natural selection, and underlying this will be changes in the DNA of the two populations. In a relatively short time, the original population will have diverged into two populations whose DNA is different enough that they can no longer interbreed. It was probably a scenario something like this that caused the divergence in the great ape lineage about 7 million years ago.

The process of speciation, and all of biological evolution, can now be understood as the change and flow of DNA through time. Scientists have learned to use DNA as a "molecular clock" because of the way it changes over time. When DNA is analyzed from humans living in all locations around the Earth, it is found to be about 99.9% similar. That is, all of our differences as humans are accounted for by only about 0.1% of our DNA! This remarkable similarity in the human genome also suggests that our species has not been around for very long, and that the various races of humans appeared quite recently (more on this later).

When human DNA is compared to chimp DNA, the two are found to be 98.4% similar (or 1.6% different). This tells us first of all that chimps and humans are closely related; DNA analysis confirms what we would suspect from looking at physical characteristics: chimps are our closest relatives. How long did it take human DNA and chimp DNA to become different by 1.6%? Scientists have now been able to estimate the rate of change of DNA, and that estimate turns out to be a change of about 1% every 4.4 million years. Using this rate, the 1.6% difference in chimp and human DNA translates into about 7 million years. This suggests that chimps and humans had a common ancestor about 7 million years ago, and then something caused the population to diverge into two different lineages.

This DNA clock technique has been used to date other divergences over the last 20 million years or so (the technique is probably not valid for times much longer than this). When gorilla DNA is compared to human DNA, a difference of 2.3% is found, showing that we are less closely related to gorillas and that our lines diverged about 10 million years ago. However the technique is limited to things that we can extract DNA from—organisms that are either living or that lived fairly recently. Scientists have now been able to extract DNA from 35,000-year-old Neanderthal bones and this has revealed important information about our relationship to Neanderthals, as we will see later in Chapter Nine. For anything much older than this, it has not yet been possible to extract viable DNA. If this *was* possible, many of the open questions in the field of paleoanthropology would be resolved.

First Hominids

It's been about 7 million years since the human lineage diverged from the chimpanzee lineage, and for the first few million years of this time we have very little evidence to show us exactly what our ancient ancestors looked like. Hominid fossils older than 4 million years were not found until 1992 and very few have been found since then (see Appendix IV for a summary of hominid fossils). We do know that early hominids were probably about 4 feet tall and very ape-like. If they walked upright on two legs, it was probably awkward because they lived in the forest and spent most of their time swinging from tree branches, as their primate ancestors had done for millions of years.

The oldest hominid fossil to date is called *Sahelanthropus tchadensis*, nicknamed Toumai. It was found in the desert of southern Chad in 2001 and has been dated to between 6 and 7 million years old. Its discoverers believe that it falls on our side of the human-chimp split, but this remains controversial. Toumai is undoubtedly very close to being the last common ancestor of humans and chimps. The partial skull indicates that it had a chimp-like brain with a size of about 350 cc, but no bones below the skull have been found so it is not known whether it was bipedal.

In the year 2000, a 6-million year old hominid fossil was found in Kenya and given the name *Orrorin tugenensis*. The fossil remains consisted of a partial femur, bits of a lower jaw and a few teeth, but no skull fragments. It is thought that *O. tugenensis* stood about 4 feet tall and had small teeth that were more human-like than chimp-like. Grooves on the femur from muscle attachments suggest that it was bipedal, but the very limited evidence leaves much uncertainty about this species.

Teams working in Ethiopia between 1992 and 2001 unearthed fossil fragments of two different species that are thought to have been members of the same genus, which was given the name *Ardipithecus*. *Ardipithecus kadabba* lived about 5.5 million years ago and *Ardipithecus ramidus* dates to about 4.4 million years ago, and both species were clearly becoming bipedal. Tim White and his team worked for 15 years to reconstruct a badly crushed *ramidus* specimen, and finally in October of 2009 revealed it to the world, having dubbed it "Ardi" (*Science Magazine* 2009). Ardi, presumably female, stood about 1.2 meters (4 feet) tall, weighed in at about 50 kilograms (110 pounds), and had a small brain with a volume of 300-350 cc. She displayed a surprising mix of apelike and monkeylike features but she did not look like a chimp, making it clear that humans did not evolve from chimps; instead, chimps and humans had a common ancestor, and Ardi offers a very good glimpse of what it looked like (see Figure 8-5).

Figure 8-5
Artist's Conception of Ardipithecus ramidus

Courtesy: J.H. Matternes

Conventional theories held that hominids became bipedal because the forests disappeared due to climate change, forcing them onto the ground. However fossil plants and animals found with the two *Ardipithecus* species indicate that they were forest dwellers, and the skeletal structure of Ardi suggests that she was capable of both tree climbing and upright walking. To further weaken the conventional theories of bipedalism, living Orangutans (considered strictly non-bipedal) have recently been observed to run on two legs across thin, flexible branches in trees (Thorpe et al. 2007), which suggests that bipedalism may have originated in trees, not on the ground, and may have roots from even before the hominid divergence 7 million years ago.

These early fossils have provoked considerable debate among paleoanthropologists. Were they really hominids? That is, were they becoming bipedal or were they actually just apes that swung from trees branches? If they were becoming bipedal, did they invent it or inherit it from long before? And do they really belong to three different genera, *Sahelanthropus*, *Orrorin*, and *Ardipithecus*, or just one? Did one succeed another in a ladder-like fashion, or were they parallel, independent branches of the early hominid tree? The only thing that everyone agrees on is that there is far too little evidence to settle these questions. In time, when new finds are made, the answers may be found.

Genus *Australopithecus*

About 4 million years ago a new hominid appeared in Africa: *Australopithecus*, the southern ape. It is not known whether the first Australopiths evolved directly from *Ardipithecus*, or the two diverged from a common ancestor. The *Australopithecus* lineage was highly successful over the next 3 million years, evolving into at least six different species (see 8-7). In the end, the lineage went extinct, but somewhere along the way (about 2.5 to 3.0 million years ago) another hominid line branched off to become the genus *Homo*, eventually leading to humans, as shown in Figure 8-6.

Figure 8-6
The Chimp-Hominid Tree

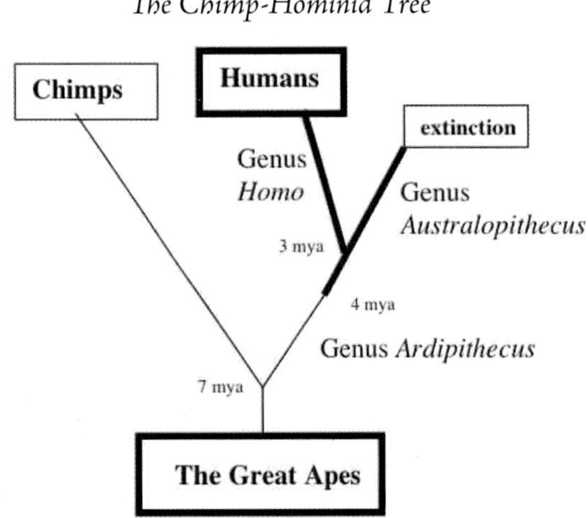

Figure 8-7
Summary of Australopithecus Members

Species	When They Lived	Brain Volume (in cubic cm)	Comments
Genus *Australopithecus*			
anamensis	4.2 – 3.9 million years ago	350 – 400	Few specimens
afarensis	3.8 – 3.0 million years ago	375 – 500	Lucy (1974) and Selam (2006)
africanus	3.0 – 2.0 million years ago	420 – 500	Likely Path to *Homo*
sediba	1.8 million years ago	420	Proposed in 2010 as precursor to *Homo.*
garhi	2.5 million years ago	550	Same as *Homo habilis*?
Genus *Paranthropus* (previously *Australopithecus*)			
aethiopicus	3.0 – 2.3 million years ago	400 – 450	Path to extinction
robustus	2.0 – 1.5 million years ago	500 – 530	Heavy- boned (robust)
boisei	2.1 – 1.1 million years ago	500 – 530	Last Australopith
Homo sapiens (for comparison)		1400 (average)	Modern Humans

From the time the hominids diverged about 7 million years ago until anatomically modern humans appeared about 200,000 years ago, three evolutionary trends are clear. The first was the gradual improvement of bipedalism, which enabled efficient movement on the ground and freed the hands for other purposes. The second trend was the increasing manual dexterity that resulted from hands with opposing thumbs and culminated with modern humans who can perform complicated surgery, create intricate paintings, or pitch a spinning baseball at 100 mph. The third trend was a steady increase in brain size, which progressed from about 350 cc in early hominids to an average of about 1400 cc in modern humans—a *quadrupling.* Although the Neanderthals had the largest brain of all hominids at 1400 to 1800 cc, they remained primitive in their culture and behavior compared to smaller-brained *Homo sapiens,* and finally went extinct. As we will see later, brain volume *alone* does not equate with intelligence. Something else happened to humans.

The earliest known member of the genus *Australopithecus* was *A. anamensis.* Although the first *anamensis* fossils were discovered in 1965 in Kenya, the species was not recognized and named until the 1990s when more finds were made. This 4 million-year-old Australopith had a sloping ape-like face, a larger brain than the earlier hominids, and stood between about 4 and 5 feet tall (with males significantly larger than females). The large, thickly enameled molars suggest a vegetarian diet of hard-to-chew foods, and the leg structure shows that they were partially bipedal. However the inner ear structure, which regulates balance on two legs, is more similar to apes than to humans, suggesting that they were still apelike in their movement and not smooth, agile runners. The fingers and toes were long and curved which was very helpful for movement in the trees. These features were typical of all of the Australopiths, summarized in Figure 8-7.

The great split in the hominid line that eventually led to humans is thought to have occurred about 3 million years ago (Figure 8-9). One widely accepted theory is that *Australopithecus afarensis* (Lucy's species) diverged into two lines that are referred to as the *gracile* (delicate boned) and the *robust* (thick boned). It is thought that the gracile line, which probably began with *A. africanus*, later gave rise to the genus *Homo* culminating with humans, while the robust line eventually went extinct. The robust Australopiths have now been given the genus name *Paranthropus*, and have sometimes been called "failed humans" because the entire line (at least 3 species) went extinct.

However, in 2010, South African paleoanthropologist Lee R. Berger suggested a different theory of the origin of *Homo* after he and his team discovered two unusual specimens of *Australopithecus* in Malapa, South Africa, dating to about 1.85 million years ago (Berger et al. 2010). Berger has proposed that this is a new species and named it *Australopithecus sediba*. Because it shares more features with early *Homo* that any other Australopith, they are proposing that *sediba* was the transitional species between *africanus* and early *Homo*. However skeptics, such as Tim White, have suggested that *sediba* is not a separate species, but a variant of *africanus*, and therefore *africanus* should retain its status as precursor to *Homo*. This debate will be settled when more specimens are discovered.

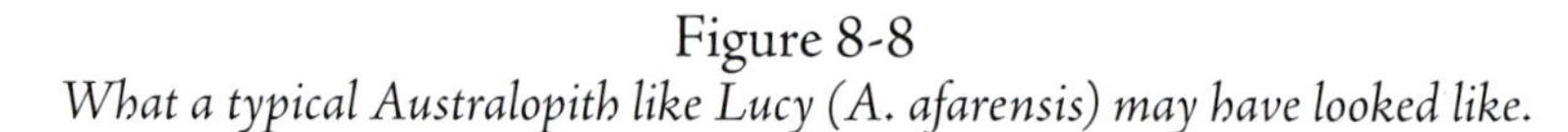

Figure 8-8
What a typical Australopith like Lucy (A. afarensis) may have looked like.

Wikimedia Commons/Cosmo Caixa, Barcelona, Spain

The question of how the genus *Australopithecus* eventually became the genus *Homo* has been complicated further after new finds announced in 2010 by a team led by Yohannes Haile-Selassie and Owen C. Lovejoy (Haile-Selassie et al. 2010). They found the skeleton of a 3.6 million year old male *afarensis* that stood an estimated 5 to 5.5 feet tall, much taller than Lucy, who stood about 3.5 feet tall. This individual, dubbed *Big Man*, had an upright stance that favored walking more than tree-climbing. Recent analysis of foot structure also indicated that Lucy's kind spent their time on the ground walking and not in trees. All of these things suggest that the species *A. afarensis* may have been much more human-like than previously thought.

All of the species of Australopithecus and Paranthropus differed from the later species in the Homo lineage in three important ways: they never left Africa, they never developed stone tools (at least, none have ever been found*), and their brains never grew beyond about 530 cc. The story of human evolution from here on will contain themes around these three things: the migrations out of Africa, the development of technology, and the growth of the brain.

Figure 8-9
The Australopithecus-Homo Split

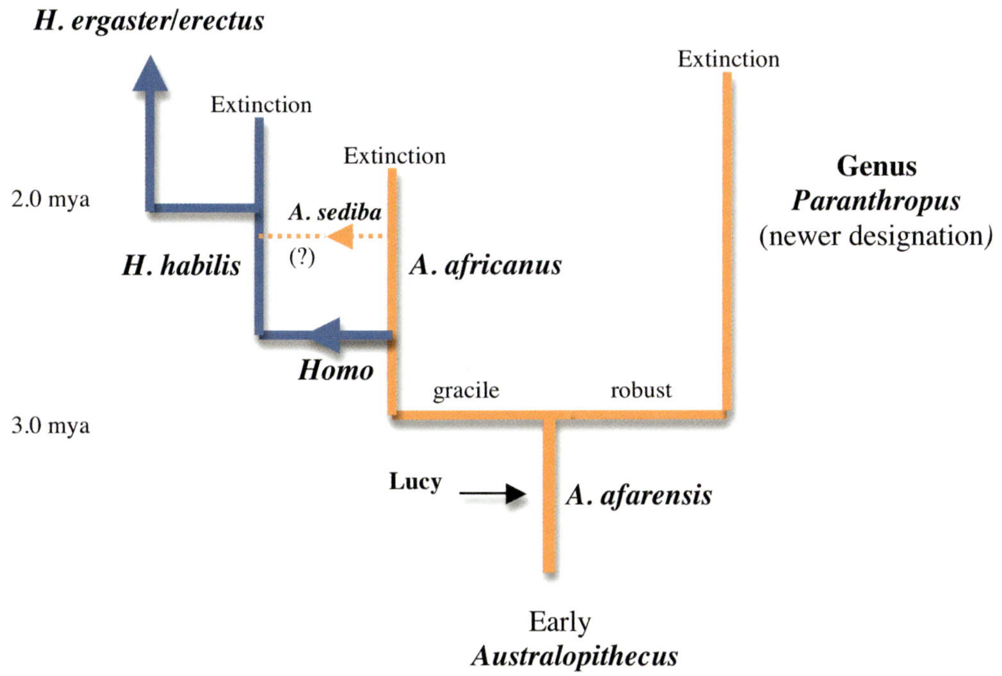

* Animal bones with scrape marks dating to 3.4 million years ago have recently been reported by Shannon P. McPherron, et al., of the Max Planck Institute for Evolutionary Anthropology in the August 12, 2010 issue of *Nature*. The authors claim that this is the first evidence of stone tool use, about 800,000 years earlier than previously thought, and suggests that *A. afarensis* was a meat eater. See also Wong, Kate, The first butchers, *Scientific American*, October 2010.

Chapter Nine
Homo: A Whole New Animal

Tool Time

About 2.5 million years ago some of the hominids in Africa began making and using stone tools. No other living thing had ever done this before. These simple stone tools have been found in large numbers near the remains of a new kind of hominid with a larger brain. It has come to be called *Homo habilis,* the "handy man." This was the beginning of the genus *Homo,* the lineage that eventually led to modern humans. Stone tools were the first significant technology ever developed by a living thing on Earth, and the use of technology has obviously become one of the hallmarks of modern humanity. Very simple tool use can be seen in a few animals, such as birds building nests out of twigs and mud, or chimps using sticks to extract termites. However only members of the *Homo* lineage had the intelligence and the manual dexterity to make and use stone tools, and to pass along the craft for hundreds of thousands of years. The manufacture and use of tools, learned and transmitted from person to person, represents perhaps the first culture among living things on Earth, and from this time forward, culture would evolve steadily in the hominid lineage.

Figure 9-1
What Homo habilis may have looked like.

Wikimedia Commons/Westfälisches Museum für Archäologie

The earliest stone tools are known as *Olduwan* tools because they were first found at Olduvai Gorge in Tanzania. These were made by striking one stone with another, causing sharp flakes to chip off. The sharp-edged flakes could be used to scrape, cut, and chop, making food gathering more efficient.

The first *Homo habilis* fossils were found at Olduvai Gorge in the early 1960s by the Leakey family and others, and dated to about 1.75 million years ago, but later finds and better dating techniques show that *habilis* lived from about 2.4 to 1.6 million years ago. *Homo habilis* is widely considered to be the first member of the genus *Homo,* and apparently evolved from *Australopithecus africanus* (Figure 8-9). Some paleoanthropologists suggest a transitional species that they name *Australopithecus garhi,* and others call *A. sediba.* Another camp of paleoanthropologists give the name *Homo rudolfensis* to a very similar hominid that lived at about the same time. For others still, the names *garhi, rudolfensis, sediba,* and *habilis* all refer to the same thing, the same species. This is a typical difference of opinion between "splitters," who like to designate many species, and "lumpers," who prefer few species. Here we will use only the name *habilis* for the first members of the *Homo* line.

The placement of *habilis* in the new genus *Homo* rather than *Australopithecus* was based on its larger brain volume of about 650 cc (the Leakeys and others proposed a cutoff of 600 cc for the genus *Homo*), smaller molars, a flatter face, and the use of tools. Being the first in the Homo lineage, *habilis* was the least human of all the *Homo* species—it still had long arms and short legs, and a brain about half the size of modern humans; it probably weighed as much as 100 pounds and males were up to five feet tall.

Homo habilis lived at the same time and in the same areas of Africa as the late Australopiths *Paranthropus boisei* and *Paranthropus robustus.* All three of these hominids must have encountered each other, but no one knows how they might have interacted. They must have competed for similar food resources, but did they avoid each other, fight with each other, or cooperate at times? *Australopithecus* and *Homo* had diverged on the evolutionary tree, so viable interbreeding should not have been possible. The two lineages would become more and more separate as members of the *Homo* line became increasingly intelligent, and the Australopiths stalled out and finally went extinct.

While the climate in East Africa was drying and cooling, turning the forests into grasslands, the Australopiths developed huge molars and strong jaws so they could process grasses, roots, and other vegetable material. This narrow vegetarian diet perhaps contributed to their stagnant brain growth that maxed out at about 530 ccs. However, *habilis* became a scavenger, eating almost anything, including meat. The simple stone tools they made could be used to strip meat off of bones or scrape and cut plant material. They were probably not yet true hunters—their crude stone tools were not tools of hunting or self-defense. There is ample evidence that *habilis* was sometimes a meal for some of the large predatory animals of the time, which meant that they had not yet reached the mastery over other animals that later hominids would accomplish.

It is not coincidental that tool making, meat eating, and rapid brain growth happened together: the three were highly interrelated. From a purely Darwinian perspective, however, the phenomenal brain growth in the *Homo* line is puzzling. A large brain seems to have as many survival *disadvantages* as advantages. For one thing, the brain is a highly energy-intensive organ. In modern humans it makes up only about 2% of our body weight yet consumes 20% of our energy. Feeding a larger brain demanded more time, effort, and calories. A large brain created a further problem during childbirth. The large head increased the risk of injury or death to both mother and child. Humans partially solved this problem by giving birth earlier, when the head was smaller, but this meant that the newborn was totally helpless, creating a significant burden on the parents. These hardly sound like survival advantages.

Still, the larger brain brought greater intelligence and tool use, which made food gathering more efficient and brought a more diverse diet, including meat. The addition of meat to the diet undoubtedly contributed to the brain growth of *Homo* and its eventual success. The brain would triple in size from the time *Homo* diverged from *Australopithecus* about 2.5 million years ago to the time the Neanderthals roamed Eurasia some 250,000 years ago.

First Out of Africa: *Homo erectus*

By about 2 million years ago, another new hominid species appeared in Africa, probably the result of climate changes. The rainforests were disappearing and being replaced by grasslands and deserts. This new hominid was much more human-like than *habilis*, with slender hips, longer legs and shorter arms, very little hair, and a larger brain. We now call it *Homo ergaster*. The tall slender anatomy and dark hairless skin were adaptations for the hot, dry climate in Africa, and *ergaster* may have been the first animal to regulate body temperature by sweating.

Perhaps driven by the deteriorating environment in Africa, some of the *ergaster* population migrated north out of Africa, probably the first hominids to do so. They moved into the Middle East, up into Central Asia, and eventually all over Southeast Asia. The name *Homo erectus* has been given to members of this species who lived outside of Africa. Eventually *ergaster* in Africa and *erectus* in Asia diverged because of their geographical separation, with the Asian *erectus* lineage ultimately going extinct, and the African *ergaster* lineage later evolving into Neanderthals and modern humans (this is one widely accepted theory).

Homo ergaster was the dividing line in hominid evolution: everything before it was ape-like and everything after it was human-like. They are sometimes called archaic humans. Alan Walker, an expert on *ergaster* and *erectus* from Pennsylvania State University, describes them as "the velociraptor of the day. If you were to look one in the eyes, it might appear to be superficially human, but you wouldn't connect. You'd be prey. It had the body of an adult human but the brain of a baby (Bryson

2003)." However that brain, typically around 1000 cc, was much larger than any other hominid's brain; *ergaster* and *erectus* were probably much more intelligent than their Australopith cousins and could easily out-compete them.

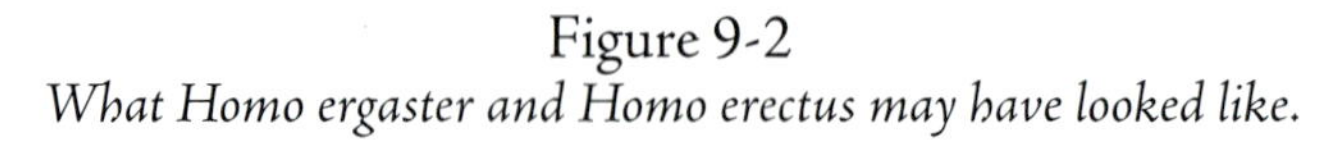

Figure 9-2
What Homo ergaster and Homo erectus may have looked like.

Wikimedia Commons/Westfälisches Museum für Archäologie

Homo ergaster in Africa and *Homo erectus* in Asia were the first to use fire, to hunt, and to care for the frail and dying. One of their most defining characteristics was the development of a new generation of stone tools that began to appear in Africa about 1.75 million years ago (Bower 2011). These stone tools were more sophisticated than the Olduwan tools of *Homo habilis*, and are called *Acheulean tools*, after the site in France where tools of this type were first found. Acheulean tools were typically hand axes (Figure 9-3). They were carefully shaped stones that fit comfortably in the hand and had a flat, protruding edge that could be used for chopping and cutting. These were so superior to the previous Olduwan flaked tools that their construction and use was passed on for nearly a million years. This suggests that very specific design criteria were handed down by explicit training within continuous proto-human societies. According to Ian Tattersall, Curator of Anthropology at the American Museum of Natural History in New York:

> They made them in the thousands. There are some places in Africa where you literally can't move without stepping on them. It's strange because they are quite intensive objects to make. It was as if they made them for the sheer pleasure of it. (Bower 2011)

Figure 9-3
Left: Acheulean hand axe from North Africa as used by Homo erectus.
Right: A modern human demonstrates its use.
This was the state-of-the art in technology for about a million years.

Courtesy: Eric D. Miller Collection.

Acheulean tools eventually showed up in Central Asia and later in Europe, but they have never been found in the Far East, even though *erectus* lived there (Java Man and Peking Man). No one understands the reason for this, but some scientists have suggested that the Far East tool culture developed around bamboo and other degradable materials that would leave no evidence.

The slender hips and large brain that made *ergaster* and *erectus* so human-like came at a cost: childbirth was more difficult. The solution to this was to give birth before the brain was fully developed, when the head was smaller. This made newborn babies helpless and vulnerable, which placed more demand on the mother for at least the first year of life. Support from the male partner became essential and this was probably the beginning of the nuclear family.

Because the *erectus* fossils called Java Man and Peking Man were found in Asia, most paleoanthropologists of the early twentieth century believed that humans originated in Asia less than a million years ago. However, the flood of hominid finds in Africa, starting in the 1960s, finally killed the Asian origin theory. The first African *ergaster/erectus* fossils were found in 1961 in Chad, and since then there have been a continual stream of African finds. In 1984 Richard Leakey and Alan Walker discovered the "Turkana Boy" in Kenya, a 90% complete skeleton of a 10-year old *ergaster* boy. The gender of the boy was determined by the pelvic structure, and the age at death by the eruption of the second molars. His projected adult size would have been about 6 feet tall and 150 pounds, with a brain volume of about 910 cc. The Turkana Boy lived about 1.6 million years ago, and other *ergaster* specimens date to about 2.0 million years ago, well before any known *erectus* in Asia, leaving little doubt that *erectus* originated in Africa and then migrated into Asia.

Recently paleoanthropologists have found hominid remains that are now classified as *erectus* at a site near Dmanisi, in the Republic of Georgia between the Caspian Sea and the Black Sea. Since 1991 a large number of fossils and tools have been collected there and dated to as early as 1.85 million years (Lordkipanidze et al. 2007). These are the oldest hominids ever found outside of Africa, suggesting a first migration out of Africa around 1.9 million years ago. Remains of African mammals have also been found near Dmanisi, showing that many animals migrated out of Africa at about this time. Climatic conditions were probably favorable for such a migration.

Homo erectus was apparently not the only hominid living in Georgia in these days. Another hominid has been found in the same layer of sediment, but this one was small-brained and short in stature, much like *H. habilis*. Because it is so different from *erectus*, it has been given a new species name, *Homo georgicus*. This now raises the possibility that *H. habilis* also migrated out of Africa and settled in Georgia. Just as *erectus* is the non-African version of *ergaster*, it may be that *georgicus* is a non-African version of *habilis*.

However *georgicus* and *habilis* both went extinct by about 1.5 million years ago, while *erectus* moved rapidly into China and Southeast Asia where it would proliferate for more than a million years. Specimens of *erectus* have been found in Java dating to about 1.7 million years ago, a relatively short time after they left Africa, and hominid fossils found in 2007 in Spain's Atapuerca Mountains suggest that *erectus* may also have moved west into Europe by about 1.2 million years ago.

The Atapuerca hominids remain controversial, with the discoverers proposing a new species *Homo antecessor* (Carbonell et al. 2008), while others consider it to be *Homo erectus* or *Homo heidelbergensis*. Whether or not *erectus* made it to Spain, there is no doubt that they were highly mobile and did get as far as China and Java. This wanderlust was probably fueled by their mastery of upright walking and running, and a meat-eating life style that supported their ongoing brain development. Carnivores generally need a much larger range than herbivores because there are fewer calories per area available to them, so *erectus* was constantly on the move. They were the first true hunter-gatherers with a nomadic lifestyle.

Homo erectus was one of the most successful of all the hominid species, with a run of nearly two million years. With a bigger brain and more sophisticated technology than any previous hominid, they spread further and adapted to a wider range of environments than any animal before. Yet in the end, *erectus* went extinct, like the Australopiths before them (there is an alternate view discussed in the next section). Fossil evidence suggests that *erectus* went into decline by about 300,000 years ago, though a few may have survived in Southeast Asia as recently as 30,000 years ago. The newly discovered fossils of dwarf hominids on the island of Flores in Indonesia (popularly called "hobbits" and tentatively given the name *Homo floresiensis*) date to about 12,000 years ago and may have been the last vestiges of *erectus*. Ultimately *Homo erectus* was not able to rise to the challenges of climate change and competition from two even more intelligent hominids: *Homo Neanderthalensis* and *Homo sapiens*.

Human Origins: Dueling Theories

The details of how modern humans originated and then spread over the entire Earth are still murky and contentious among paleoanthropologists. For the last few decades there have been two competing theories of humanity, generally called *Out of Africa* and *Multiregional Evolution* (see 9-4).

Recall that *Homo ergaster* was probably the first hominid to leave Africa about 1.9 million years ago, and by about 1.8 million years ago they showed up in Central Asia (modern Georgia) as *Homo erectus*, and as far away as Java soon after. Many paleoanthropologists, subscribing to the *Out of Africa* theory, believe that by about one million years ago, climate changes in Africa brought about a new species of hominid, descended from *Homo ergaster*. This was *Homo heidelbergensis* (the first specimen was found near Heidelberg, Germany in 1907). They were apparently the second hominid species to migrate out of Africa. Fossil evidence indicates that *heidelbergensis* reached Europe by about 800,000 years ago, and perhaps earlier. With a brain nearly as large as modern humans, *H. heidelbergensis* was significantly more intelligent than *ergaster* and *erectus*. They not only brought the traditional Acheulean tools out of Africa but also developed new and better tools like spears with tips of sharp stone. There is some evidence that *heidelbergensis* buried their dead, making them the first hominid to do so, and used the red pigment ochre for art and ritual. These are signs of more advanced culture, surely driven by a more sophisticated functioning of the brain.

The Out of Africa theory further asserts that while some members of the *heidelbergensis* population migrated out of Africa, others remained in Africa, dividing the population into two. In a classic example of *allopatric speciation*, the two populations separated genetically. The northern population in Europe faced a series of ice ages with times of harsh cold. They adapted to this by developing a shorter and stockier frame for conserving body heat, wider nostrils that heated the cold air, and a thick skull with heavy brows that protected the brain from the cold. They became *Homo neanderthalensis*, the Neanderthals, by perhaps 400,000 years ago. Meanwhile the African population of *heidelbergensis* retained their tall, slender frame that was better suited for a hot climate and finally evolved into *Homo sapiens*, anatomically modern humans, by about 200,000 years ago.

In the Out of Africa scenario, also known as the *Replacement theory*, modern humans then left Africa between about 100,000 and 50,000 years ago, and eventually spread all over the Earth, replacing the two other hominid species *erectus* and *neanderthalensis* who both eventually went extinct. The different populations of modern humans (such as Asians, Africans, and Europeans) would all have appeared quite recently, in the last 50,000 to 100,000 years or so, and would be very closely related. If we all came out of Africa in the last 100,000 years, the DNA of all living people should be very similar.

The alternate theory of Multiregional Evolution, whose chief proponent has been Milford Wolpoff from the University of Michigan, holds that after *Homo erec-*

tus spread all over Africa, Asia, and Europe by about 1.5 million years ago, modern humans gradually evolved from these dispersed populations of *erectus*. In this view, *erectus* did not go extinct but gradually evolved into modern humans in many regions of the old world—Africa, Europe, Asia, and Australia—and these regional varieties became the races of humans we find today. This theory has been incorrectly interpreted as suggesting that some kind of independent parallel evolution of humans occurred in different regions of the world, but instead, it proposes a worldwide network of genetic exchanges. That is, it assumes that humans were migrating and breeding across long distances, sharing both genes and culture (this is represented by the horizontal lines in Figure 9-4). Strict multiregionalists do not recognize other species designations such as *Homo ergaster, Homo heidelbergensis,* or *Homo neanderthalensis* as separate species, but view these as transitional forms in the gradual morphing of *Homo erectus* into *Homo sapiens*. An important prediction of the Multiregional Theory is that the modern populations (races*)of humans would have separated long ago—perhaps a million years ago—and this should be reflected in significant genetic dissimilarities when DNA is analyzed from living people of all populations.

Figure 9-4
The Replacement (or Out of Africa) Model versus the Regional Continuity (or Multiregional) Model

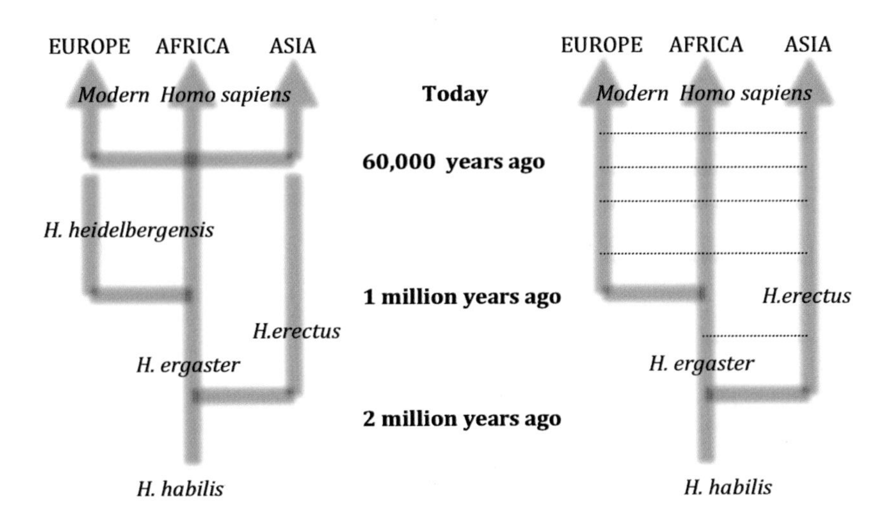

Today the evidence has mounted overwhelmingly in favor of the Out of Africa theory. Genetic comparisons of living humans show that we are all extremely similar, regardless of population type or where we live. This means that the different populations of living humans must have diverged very recently. DNA analysis suggests that

* Many evolutionary scientists have now rejected the term *race* in favor of *population, ethnicity, cline, clade, variety, or subspecies*. Most differences among humans today are not so much genetic as cultural, social, historic, and geographic, and with widespread intermixing, the notion of race is simply not very accurate or appropriate. Henceforth we will use the term *population* instead of *race*.

all humans can be traced to a common ancestor in Africa about 150,000 years ago and that the different population types diverged only in the last 60,000 to 100,000 years (more on this in Chapter 10), far different from the million years or so that Multiregionalism predicts.

Another genetic trend also seems to target Africa as our recent place of origin. A study published in 2008 (Feldman et al.) analyzed the genes of 938 individuals from 51 different populations living around the world and found that African populations today exhibit more genetic diversity than any other populations on Earth (suggesting that they are the oldest) while genetic diversity *decreases* with distance away from Africa. This trend was also found in a study published in 2007 (Hanihara et al.) that compared human skulls rather than human genes. Teams from Japan and England studied over six thousand skulls from worldwide populations, making 37 precise measurements on each skull. When they analyzed this vast amount of data, they found that the diversity was greatest in African skulls and then decreased steadily with distance from that continent. Because the evidence now so strongly supports the Out of Africa theory, our story will continue to recount this scenario.

Science and Discovery

The Mystery of Atapuerca

An alternative view of hominid evolution in Europe has recently emerged after spectacular discoveries in the limestone caves of the Atapuerca Mountains in Northern Spain. Starting in 1994 Spanish paleontologists (Rose 1997) began to find the remains of a hominid that shared features of *ergaster, erectus,* and modern humans. They proposed the new species name *Homo antecessor* for this hominid because they believed it did not fall under any of the known species. The first finds were dated to about 800,000 year ago and by 2007 the ongoing excavations had yielded fragments dating to 1.1-1.2 million years old, long before any hominids were thought to have lived in western Europe, raising the questions of who these hominids were and how did they get to Spain a million years ago?

In the mainstream view, *heidelbergensis* left Africa by perhaps 900,000 years ago to become the first hominids in Western Europe, showing up in Spain and England by 800,000 years ago. They eventually evolved into the Neanderthals. Yet now we find clear hominid presence in Spain by about 1.1 million years ago, long before *heidelbergensis* is thought to have arrived. So who were the Atapuerca hominids?

The paleoanthropologists who are excavating the Atapuerca sites have suggested an alternative scenario that we will call the Spanish hypothesis. In this view, the African population of *ergaster* evolved into *Homo antecessor* by perhaps 1.3 million years ago. The *antecessor* population then split, with some remaining in Africa to eventually become *Homo sapiens* and others leaving Africa and entering Europe, where they arrived in Spain by about 1.2 million years ago. The European *antecessor* population then evolved into *heidelbergensis* and finally into *neanderthalensis*. The Spanish hypothesis asserts that *antecessor* was the common ancestor of humans and Neanderthals, while in the mainstream view *heidelbergensis* was the common ancestor of the two (Figure 9-5, next page).

Many paleontologists do not recognize *antecessor* as a separate species, or subscribe to the Spanish hypothesis of European evolution. How, then, can they explain the Atapuerca hominids? The simplest explanation may be that they were a variety of *Homo erectus*. Recall that by 1.8 million years ago *erectus* had left Africa and arrived in Central Asia, and then moved east into China and Southeast Asia. Why wouldn't some have headed west into Europe? The problem with this scenario is that few *erectus*-like fossils have been found in Western Europe, and none dating from between 1.7 and 1.2 million years ago. The evidence is simply not there yet to determine whether or not the Atapuerca hominids are a modified *Homo erectus* or an entirely separate species. An even more controversial possibility is that some members of the African *ergaster* population were able to cross the Mediterranean and colonize Spain. Whoever these first Europeans turn out to be, they were apparently displaced by *heidelbergensis* who was probably more intelligent and competitively superior. New finds are still being made in Spain, but for now the mystery of the Atapuerca hominids remains unsolved.

Figure 9-5
Two Views of Hominid Evolution in Europe

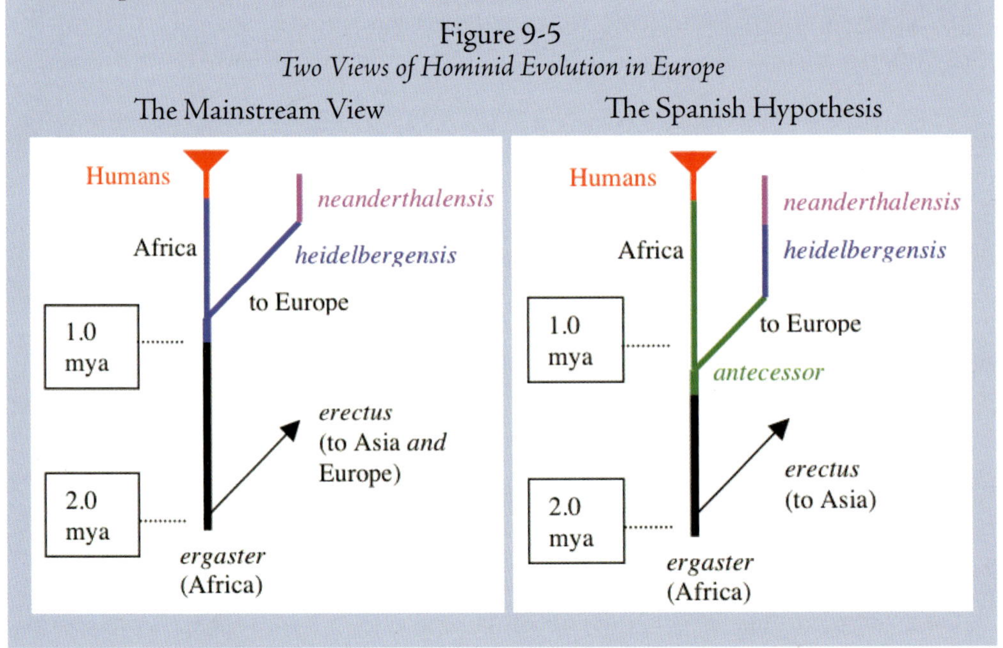

The Neanderthals

Neanderthal remains were first found in Belgium in 1829 and have since been found from Southern Siberia to Gibraltar and as far south as Israel. They were not recognized as ancient hominids or given the name *Homo neanderthalensis* until after the 1856 discovery in the Neander Valley of Germany. Subsequent finds show that Neanderthals lived all over Europe, the Middle East, and Central Asia from about 400,000 years ago until about 30,000 years ago, when they completely disappeared.

Unfortunately, Neanderthals got a bad reputation in the early 1900s because of the work of the French paleontologist Marcellin Boule. He described them as dull-witted, brutish, ape-like creatures that walked hunched over with a shuffling gait—the stereotypical caveman. He formulated this picture after studying the nearly complete skeleton of a Neanderthal excavated in 1908 at La Chapelle-aux-Saints in southwest France. This view persisted into the 1950s until further study of the skeleton showed that this was an old man with severe arthritis in his spine, giving him a hunched-over posture. As more artifacts and remains have been uncovered, the Neanderthals are now being seen as much more intelligent and human-like than originally thought.

Today we have the remains from more that 400 Neanderthal individuals, more than any other hominid—yet they are still puzzling and controversial. Were they our ancestors? That is, did modern humans evolve from Neanderthals, or were they a parallel species, with whom we share a common ancestor? Why did they vanish, or did some survive and become part of the modern human population? And what was the relationship between early modern humans and Neanderthals, who coexisted off and on for about 70,000 years? Did they interbreed?

The answers to some of these questions are now emerging as researchers have begun to extract and analyze DNA from Neanderthal bones. In 1997 a German team led by Svante Pääbo from the Max Planck Institute for Evolutionary Anthropology made the first successful extraction of DNA using a bone from the 1856 Neanderthal Man (Krings et al. 1999). They were able to analyze mitochondrial DNA and found that it was different enough from any known human DNA to suggest that humans and Neanderthals were quite different, but parallel, species that had split apart hundreds of thousands of years ago. These recent analyses suggest that humans and Neanderthals diverged at least 370,000 years ago, much earlier than the 150,000 years ago derived from the very sparse fossil record.

Figure 9-6
What a Neanderthal may have looked like.

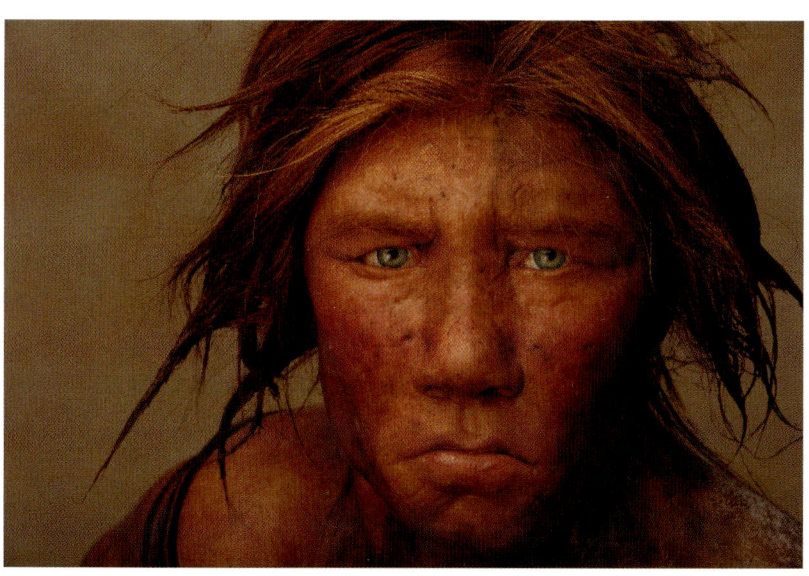

Courtesy: Joe McNally

Some scientists questioned this first published finding because DNA extraction and analysis is so difficult. Contamination from microbes near the bones or from the human handlers would be almost impossible to avoid. Since the first report, however, other teams have extracted DNA from different Neanderthal specimens using the utmost care to prevent contamination, and they have analyzed both mitochondrial and nuclear DNA using different techniques. In 2006, teams led by Svante Pääbo from the Max Planck Institute and Edward Rubin from the Joint Genome Institute in California found that the genomes of modern humans and Neanderthals were 99.5% identical, compared to the genomes of all living humans that are 99.9% identical. Based on this 0.5% difference, they concluded that modern humans and Neanderthals shared a common ancestor (presumably *Homo heidelbergensis*) about 700,000 years ago, and that the two split into separate species about 370,000 years ago. This is depicted in 9-7, from the November 2006 paper that appeared in the journal *Science* (Pääbo et al.). These findings suggested that as distinctly separate species there was no possibility of viable interbreeding between humans and Neanderthals, even though they had periodic contact for some 70,000 years and lived in very close proximity in Europe for nearly 15,000 years. They apparently shared knowledge of tools and primitive culture, but not genes.

However this view was shaken in 2010 when Pääbo and his team completed an even more extensive analysis of nuclear DNA from three Neanderthal bone fragments found in a Croatian cave (Pääbo et al. 2010). They compared the Neanderthal genomes to those of five living humans from locations around the world—from China, France, Papua New Guinea, Western Africa, and Southern Africa. While the research team was expecting further confirmation that there was no gene flow between humans and Neanderthals, they found just the opposite. The people from Europe and Asia had inherited between 1 and 4 percent of their DNA from Neanderthals! There were no traces of Neanderthal heritage in the two Africans studied, which supports the view that Neanderthals never lived in Africa. So, now it seems that Neanderthals and humans must have interbred somewhere along the way in Europe and Asia. In Pääbo's words, "Neanderthals are not totally extinct; they live on in some of us" (Saey 2010). However if Neanderthals and humans did interbreed, should they be considered separate hominid species? Further research will shed light on this and other questions about who the Neanderthals were.

Pääbo and his research team complicated the picture of hominid evolution in Eurasia even more in 2008 when they extracted and analyzed mitochondrial DNA from a finger bone and a tooth unearthed at Denisova Cave, Siberia, dating to between 30,000 and 48,000 years ago (Reich et al. 2010). Analysis of the mtDNA showed that it was significantly different from that of Neanderthals or humans, so apparently there were three different forms of hominids living in Siberia at this time. The research team will next look at nuclear DNA from the Denisova hominid in hopes of finding out more about who they were and what their relationship was to the other two hominids that shared that part of the world.

Figure 9-7
The Divergence of Neanderthals and Humans

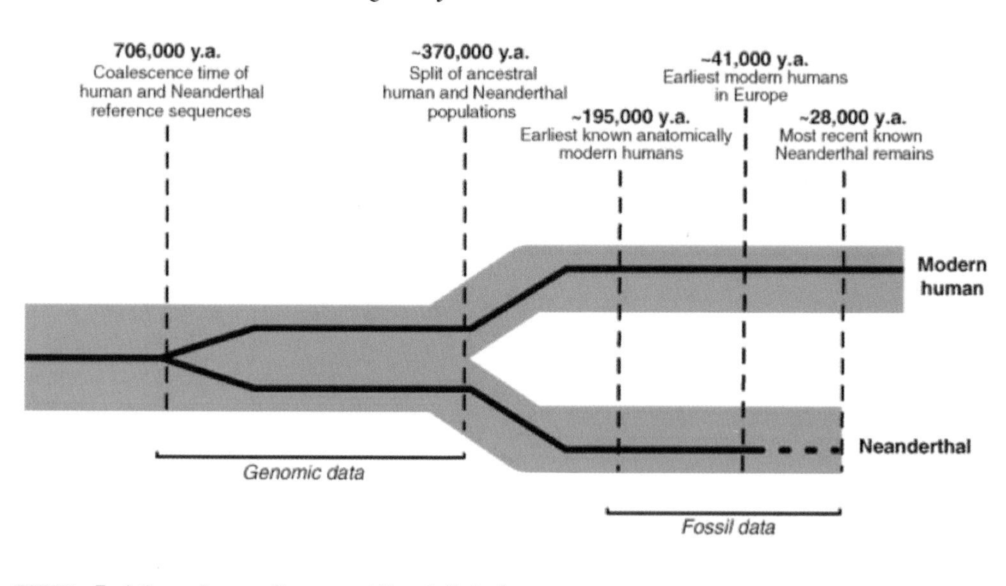

From Pääbo, Pritchard, and Rubin. Reprinted with permission from AAAS

Although little is known about the as yet unnamed Denisova hominid (*Homo denisovensis?*), the Neanderthals have been studied extensively. It is believed that Neanderthals had developed significant social structure while living in clans. As part of their social network, they hunted in cooperative groups and were probably capable of speech and simple language. In 2007 Pääbo and his team announced that they had found the FOXP2 gene in Neanderthal DNA (Pääbo et al. 2007). This gene has been strongly linked with speech and language in humans, so it seems that Neanderthals may have had this ability to some extent. Their brains were, as far as we can tell, structurally identical to ours but about 12% larger (about the same as ours when adjusted for body mass). Their tools, called Mousterian technology, were more elaborate and specialized than any before them, and included better spears for hunting, sharp blades for slicing, serrated blades for sawing, and awls for making holes. Later Neanderthals, perhaps sharing knowledge with modern humans, developed fishhooks and harpoons using bone and wood as well as stone. They cared for the sick and elderly and buried their dead, they used pigments and made jewelry, all signs of significant intelligence. There is evidence that they made protective clothing and shelters out of animal skins and by about 60,000 years ago, they may have been making simple flutes, the first known musical instruments.

Neanderthals endured some of the harshest conditions in hominid history during the last ice age, and had a life expectancy of only about 40 years. The climate in Europe varied drastically from arctic-like conditions to periods of relative warmth. In the coldest times, there were ice sheets thousands of feet thick covering England and Northern Europe, so they had to move south where they could hunt in forests. The total size of the Neanderthal population was probably never very large and appears to have diminished steadily beginning around 35,000 years ago until they disappeared altogether by about 30,000 years ago. It is not clear what happened to them, but their relatively abrupt disappearance coincides with the arrival of modern humans in Europe.

It seems likely that Neanderthals could not compete effectively with the technologically and socially more advanced humans and their burgeoning population. Contributing to the demise of the Neanderthals was probably the deteriorating climate, as Europe moved into the coldest phase of the last ice age. Perhaps their technology could not protect them from the extreme cold, while the humans may have developed better clothing and shelter. Sewing needles made out of bone have been found in human settlements, suggesting that they may have made very effective clothing and tents, while sewing needles have never been found among Neanderthal artifacts. Humans also had a very diverse diet and they could adapt to a wide range of conditions, while Neanderthals relied heavily on hunting large mammals whose supply was

Figure 9-8
Summary of the First Hominid Migrations Out of Africa
(according to the mainstream view)

Species	Appeared in Africa (years ago)	Left Africa (years ago)	Where They Went
Homo ergaster	2 million	1.9 million	Asia, and maybe Europe, as *Homo erectus.*
H. heidelbergensis	1.2 million (?)	1.1 million (?)	Middle East and Europe, becoming the Neanderthals.
---------- Looking Ahead to Chapter 10 ----------			
H. sapiens idaltu (anatomically modern humans)	200,000	100,000	The Levant, Arabian Peninsula, perhaps India and China. Faced the eruption of Mount Toba.
H. sapiens sapiens (behaviorally modern humans)	70,000 (?)	60,000	Everywhere (eventually)

quite variable. Adding to this may have been an inherent inefficiency in the Neanderthal biomechanics. Hominid energetics experts have estimated that the Neanderthals' heavy build and short shin bones would require about 32% more calories than the lighter, more nimble humans, an expensive demand in lean times (Wong 2009).

It seems that humans may simply have had many small advantages that gave them the survival edge over Neanderthals during a time of highly unstable climatic conditions. By about 30,000 years ago the Neanderthals in Europe had retreated as far south and west as possible, into warmer areas. The last known Neanderthal sites are in modern Portugal and date to about 28,000 years ago. They had their time of dominance for several hundred thousand years, but in the end the last hominid standing was *Homo sapiens*.

Science and Discovery

Climate Variability and Hominid Adaptability

One of the most important factors that made Earth an incubator for complex life is that conditions near its surface have been relatively stable for billions of years. This allowed life not only to survive, but also gave it time to evolve into the many complex forms we have today, including humans. However, within this window of stable conditions, there have been significant smaller changes that many species of life could not survive, including the great oxidation catastrophe, the snowball events, wide swings in atmospheric CO_2 and O_2 levels, volcanic winters, asteroid impacts, magnetic field reversals, and regular cycles of climate change.

One of the ways scientists can study climate change is by looking at the levels of oxygen isotopes in the past. Most of the oxygen on Earth is of two types (isotopes): about 99% is ^{16}O (with 8 protons and 8 neutrons), and most of the rest is ^{18}O (with 8 protons and 10 neutrons). Both of these combine with hydrogen to make water, but water made from ^{18}O is slightly heavier and behaves slightly differently when it comes to evaporation and condensation. The heavier water containing ^{18}O does not evaporate as easily, but condenses more readily than light water. When water evaporates from the ocean, more ^{18}O is left behind and more ^{16}O will be found in the atmosphere.

When the temperature of Earth is higher, there is more evaporation and therefore more ^{18}O will be found in the oceans and less in the atmosphere. That is, the oceans will have a higher ^{18}O to ^{16}O ratio, and the atmosphere will have a lower ratio. Microorganisms in the ocean that take in oxygen will therefore also have a greater proportion of ^{18}O. When shell-forming microorganisms die and settle to the sea floor, limestone is formed, and in that limestone the ^{18}O levels are preserved from the time the organism was alive. Scientists can drill core samples in limestone sediments and look at the ^{18}O record from millions of years in the past and, from this, get a temperature record.

More recent climatic records are preserved in core samples taken from polar ice. This ice contains air that was trapped as snow fell and then formed ice as year after year of snow piled up, and this trapped air can be analyzed to find CO_2 and methane levels, which strongly mirror Earth's temperature. Oxygen isotope ratios can also be measured in the ice to give another record of Earth temperature that extends back hundreds of thousands of years. From both limestone core samples and ice core samples climate scientists have been able to construct extensive temperature records, as shown in Figures 9-9(a) and (b) below.

Figures 9-9

(a) Temperature variations over the last 5 million years derived from oxygen isotope ratios in limestone core samples (Lisiecki and Raymo 2005).

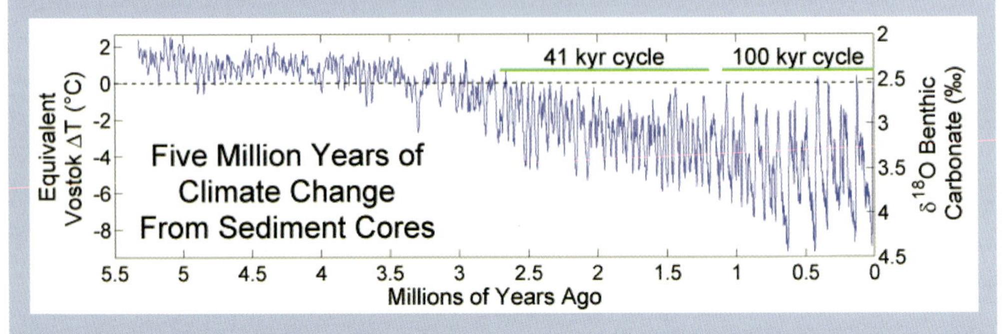

From Lisiecki and Raymo, *Paleoceanography* (2005)

(b) Temperature variations over the last 2000 years based on ten different sets of data

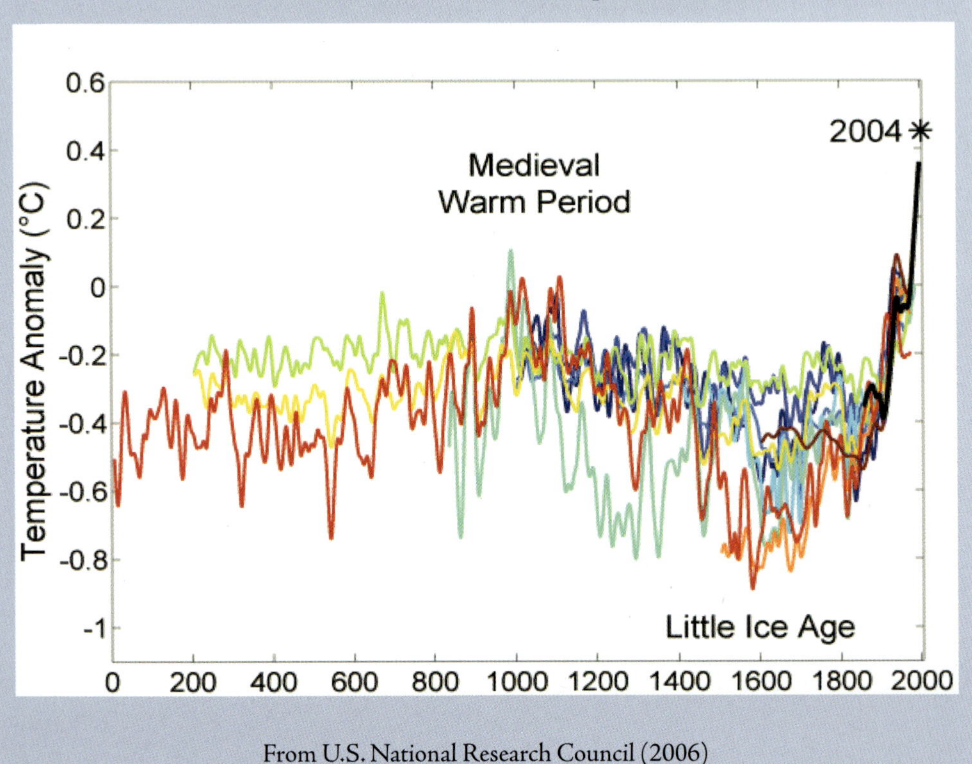

From U.S. National Research Council (2006)

One of the most important things this temperature record shows is a very significant increase in climate *variability* over the last 2.5 million years, when hominids were evolving in Africa. Such variations can spell extinction for any species that is unable to adapt to drastic changes in temperature, plant cover, and food supply; and indeed none of the Australopiths did survive this period; yet the *Homo* lineage was able to. By around 1.7 million years ago *Homo erectus* had moved into such diverse habitats as Central Asia (Georgia) with mountains, grasslands, and forests; China, with an arid climate and large seasonal temperature changes; and Java, with a tropical, coastal environment. They made tools from lava in Georgia, stone in China, and probably bamboo in Java. This was a highly adaptable species.

Conventional evolutionary theory holds that organisms are successful if they can adapt to specific changes in their habitat—this is known as a *habitat-specific hypothesis*. As an example, the savannah hypothesis of hominid evolution states that many important hominid adaptations, such as upright walking and tool-making, were driven by the loss of forest to an expanding arid savannah. If this were true, hominid fossils with those adaptations would be found only in specific environments, yet they are found widely across diversely different environments.

An alternative theory of hominid evolution was proposed by Richard Potts, curator of anthropology at the Smithsonian National Museum (Potts 1996). The *variability selection hypothesis* asserts that hominids were so successful not because they adapted to any one type of habitat but because they could quickly adapt to such a wide and variable range of habitats. They adapted to environmental *instability*. How was the *Homo* lineage able to do this while the Australopiths, and virtually all other living things, could not? The following unique suite of abilities contributed to our great success:

- Upright walking allowed hominids to move out of the trees and onto the ground. Australopiths such as Lucy still had long arms and long grasping fingers so that moving around in trees was still possible. Upright walking also freed the hands, which later made tool construction and use possible.
- An increasing brain size enhanced problem solving ability so that new solutions to survival problems could be found.
- Tool use gave access to a much more diverse range of food sources, including meat. Tool use was the product of a larger brain, the freed up hands, and unprecedented manual dexterity.
- Later hominids like *heidelbergensis* developed increasingly sophisticated tools with specific purposes, enabling them to hunt and fish effectively, and eventually to construct clothing and shelters.
- As brain size grew, so did social networking and long distance exchange of goods. Humans and Neanderthals both collaborated by hunting in packs and sharing knowledge, but humans eventually surpassed the Neanderthals as symbolic thought and language blossomed. Advanced hominids were capable of learning from each other: this was the basis of culture; and human culture could evolve rapidly.

The story of hominid evolution, then, is very much about our ability to adapt quickly to a wide range of environments. Despite such success, every hominid species must eventually have reached its limit of adaptability— except for *Homo sapiens*, for we are the only ones that did survive. Yet our greatest challenge may still lie ahead. Will we be able to adapt to the massive changes in our environment that we ourselves are bringing about?

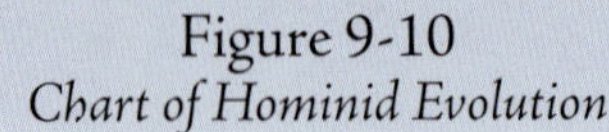

Figure 9-10
Chart of Hominid Evolution

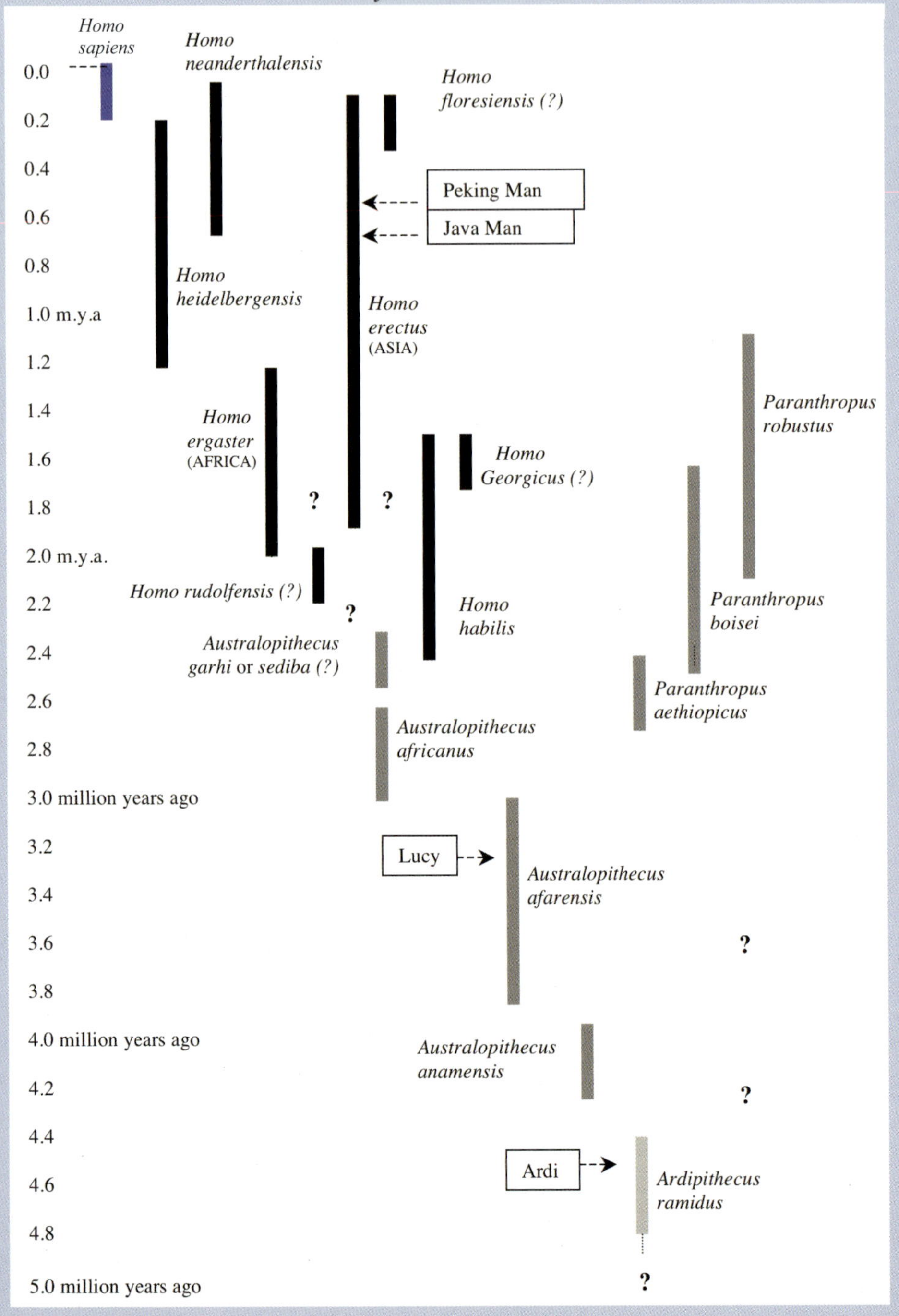

Part Four Summary

When the great apes were flourishing in Africa about 20 million years ago, it was 6 minutes before midnight on the third day in the life of the universe. Life on Earth had been evolving for 3.8 billion years, but the most dramatic development was still to come: the emergence of humanity, a life form that would become a geologic force.

By 16 million years ago, orangutans appear in the fossil record, marking the beginning of a new branch of the evolutionary tree. Then about 9 million years ago the gorillas branched off, and finally about 7 million years ago the lineage of the great apes split one more time. This split, however, had enormous consequences. One side of the split became the chimpanzees and bonobos, and the other side became a new kind of ape that began walking upright on two legs and using their freed up hands. These were the hominids. The hominid lineage produced mostly failures, or at least nearly every species in the lineage eventually went extinct. In the end only one hominid species has survived and that one species—*Homo sapiens*—became more powerful and influential than anything that had ever lived on Earth before.

The first three million years of hominid evolution left scant evidence for us to study today. The few skeletal remains that have been discovered recently suggest an ape-like creature that was beginning to walk on two legs instead of four, with a brain volume of about 350-400 cc, on par with today's chimpanzees. By about 4 million years ago the fossil record shows that a slightly more human-like form was emerging: *Australopithecus*. Over the next 3 million years *Australopithecus* would evolve into at least seven different species in Africa as the brain volume gradually grew to about 500 cc. However this is where brain size stalled out for the Australopiths, and eventually they all went extinct.

Along the way, about 2.5 million years ago, the hominid lineage split, giving rise to the genus *Homo*. This split had spectacular consequences: the Australopith side of the split would not survive, and most of the *Homo* side would not survive either, but the one species that did survive lives today as modern humans. While the brain growth of the Australopiths stalled out at about 500 cc, the *Homo* brain grew steadily, eventually reaching volumes of 1300 to 1800 cc. The Australopiths remained more ape-like than human-like, while the members of *Homo* got taller, more slender, more upright, and developed a graceful gait and smooth skin.

The first member of the *Homo* lineage is thought to be *Homo habilis*, though our knowledge of this species is very limited. Two things distinguished *habilis* from its Australopith cousins: a larger brain and the first use of stone tools. From here on, tool use became one of the most important themes in the evolution of the *Homo* lineage. As the brain of each new species of *Homo* got bigger, the tools become more sophisticated and powerful.

For almost 40 million years, the story of primates evolving into the genus *Homo* takes place entirely in Africa. Finally about 2 million years ago a new species of *Homo* evolved that was the able to leave Africa. *Homo erectus* quickly spread over much of

Asia and Europe, getting as far west as Spain and as far southeast as Java. With the largest brain yet (growing to 1200 cc) *erectus* developed effective tools for hunting, mastered the use of fire, and began living in nuclear families. They were capable of killing any other living thing for more than a million years, but they too would eventually die out. While *erectus* was wandering all over Eurasia, another part of the *Homo* lineage was evolving into a new species back in Africa, and this new species had an even larger brain: this was *Homo heidelbergensis*.

By perhaps 1.2 million years ago *heidelbergensis* appeared in Africa, probably derived from *Homo ergaster/erectus*. Some of the population migrated out of Africa and reached Europe by perhaps one million years ago, while some members remained in Africa. The two groups diverged into two separate species because the two environments were so different. The European population, facing much colder conditions, became *Homo neanderthalensis* by about 400,000 years ago, while the African population slowly evolved into *Homo sapiens*. Yet it was the Neanderthals who proliferated for the next 350,000 years, spreading all over Central Asia, Europe, and the Middle East, and probably displacing *erectus* in these areas. They had bigger brains than even modern humans, they were strong and could endure brutally cold conditions, and their tools were more sophisticated than anything before. Yet just like every hominid species before them, they too would eventually go extinct.

Meanwhile back in Africa, by about 200,000 years ago, the *heidelbergensis* population had evolved into anatomically modern humans, *Homo sapiens*. These first people looked physically very much like today's humans, but they were still primitive in their intelligence, culture, and behavior. Many challenges lay ahead for them, including near-extinction, before they could surpass the Neanderthals and spread over the Earth, and that is the subject of the last part of our story.

Part Five

The Age of Humanity

PART FIVE

THE AGE OF HUMANITY
Last Hominid Standing

Our story so far has spanned almost 14 billion years, from the birth of the universe to the reign of the Neanderthals. This covers all but the last few seconds of the three geocosmic days that represent the life of the universe. It was in those last seconds that full-fledged human beings emerged and acquired powers far greater than anything that had ever lived on Earth.

By about 250,000 years ago, the Neanderthals lived all over Eurasia and were the most intelligent and powerful animal on these continents. They were large-brained, physically powerful, cold-adapted, and very human-like; their stone and bone technology was more advanced than anything before. Yet over the next 200,000 years, the fossil record shows that their tools hardly advanced; the Neanderthals seem to have stalled out in their development.

Far away on the continent of Africa, a parallel hominid species was emerging by about 200,000 years ago: *Homo sapiens*. Unlike the Neanderthals, modern humans did anything but stall out. From our modest beginnings in Africa we humans have steadily evolved in our intelligence and culture, right up to the present day. This was not physical evolution but something new, something inside, associated with the brain: an advancing cognitive ability, the emergence of language and culture, and what some have called self consciousness. In a now widely recognized "great leap" that was probably unfolding around 60,000 years ago, humans acquired a whole new suite of abilities unlike any living thing before. This dramatic rise in intelligence, this expansion of consciousness and culture, launched us far beyond the Neanderthals to make us competitively superior by almost any measure. As people spread over Earth, *Homo neanderthalensis* and the last vestiges of *Homo erectus* disappeared.

Out of the entire hominid lineage—from *Ardipithecus*, to the many species of *Australopithecus*, to all the versions of *Homo*—we are the only one that survived; we are the last hominid standing. Modern humans were capable of much more than just surviving, and by about 5000 years ago we had created something never before seen on Earth: *civilization*.

Time Context for Part Five

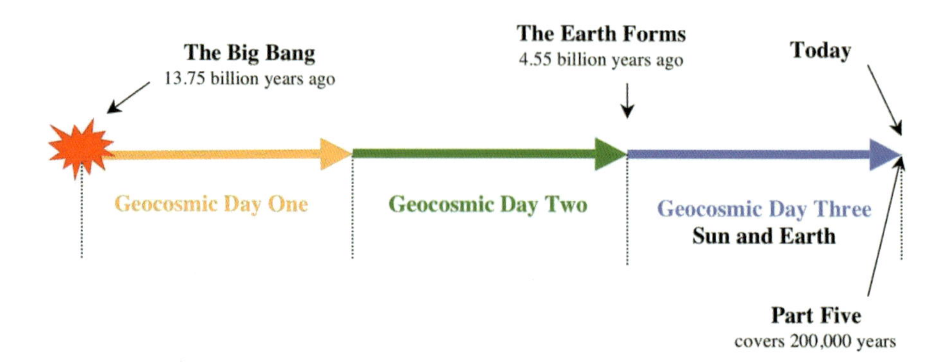

Main Events of Part Five: The Age of Humanity
All dates are approximate

Event	Years Ago	Geocosmic Time (all are Day 3)
Anatomically modern humans appear in Africa	200,000	11:59:56 pm
1st attempts by humans to leave Africa	100,000	11:59:58
Eruption of Mt. Toba and the near-extinction of humans	73,500	
The "great leap" is completed: behaviorally modern humans emerge and successfully leave Africa	60,000	11:59:59
People arrive in Australia	50,000 (?)	
People move into Europe	40,000	
Last Neanderthals (southwestern Europe)	28,000	11:59:59.5
People leave Asia and move into Alaska	20,000–16,000	
People reach Patagonia and now cover the Earth	14,000	
End of last Ice Age, beginning of Neolithic Period	11,500	
First agriculture and proto-cities (Jericho)	10,000	11:59:59.8
First city-states, hierarchical societies, and writing: formal history and civilization begins	5,200	11:59:59.9

Chapter 10
Idaltu: Facing Extinction

The First *Homo sapiens*

While the Neanderthals were flourishing in Europe and Central Asia several hundred thousand years ago, another hominid was evolving in Africa along a different path. The members of *Homo heidelbergensis* who had remained in Africa (recall that some left Africa about a million years ago to become the Neanderthals) were tall and slender, with finer features, a lighter build, and a large brain. They were slowly coming to resemble modern humans. Multiple lines of evidence now suggest that *Homo sapiens*—the symbol user—emerged sometime between 250,000 and 200,000 years ago in Eastern Africa, in or near modern day Ethiopia.

Fossil evidence of the earliest *Homo sapiens* dates to almost 200,000 years ago. This evidence consists of the partial skeletons of two early humans that were found in 1967 by Richard Leakey's team in Kibish, Ethiopia. They were dubbed Omo I and Omo II and originally dated to be 130,000 years old, but better dating techniques used in 2007 moved their origin back to 195,000 years ago. These are the first true humans that we have fossil evidence of and are referred to as *anatomically modern humans*. The next oldest specimens of *Homo sapiens* are 160,000 years old and were found in Herto, Ethiopia by Tim White's team in 2003.

These early African *Homo sapiens* are sometimes given the subspecies name *Homo sapiens idaltu* because they had remnants of archaic features, such as a slightly heavier build, larger molars, and more pronounced brow ridges than we have today. Over the last 200,000 years humans have continued to evolve a lighter bone structure and smaller molars, but our brain size has remained roughly the same. Though *idaltu* was very similar to us *anatomically*, its large brain probably functioned at a relatively primitive level compared to modern day humans, perhaps not even at the level of the Neanderthals. Over the next 200,000 years this full-sized brain would gradually take on the higher cognitive functioning and self-aware consciousness that is the unique hallmark of *behaviorally* modern humans, *Homo sapiens sapiens*.

The scant fossil evidence points to Africa as the birthplace of modern humans some 200,000 years ago, and this is supported more recently by another line of evidence, embedded in the DNA of living humans. In 1987 Allan Wilson and his students Rebecca Cann and Mark Stoneking published a groundbreaking paper based on the analysis of mitochondrial DNA (mtDNA) from living humans (Cann et al. 1987). After analyzing mtDNA samples from 147 different people from five geographic regions, Wilson and his team concluded that all humans living today are descended from a single woman who lived in Africa about 200,000 years ago.

After this conclusion was published, the press soon touted this ancestral woman as the "Mitochondrial Eve," the "African Eve," and the "Black Eve," and many misconceptions have swirled around the idea ever since, including the notion that she was the first woman, or the Biblical Eve (see Science and Discovery). More recently, Y-chromosome studies of living males indicate that all living males are descended from a single man in Africa who lived about 60,000 years ago—the "Y chromosome Adam" (Chapter 11).

Y chromosome is passed from father to son.

We inherit only our mother's DNA and thru female lineage. We all come from one woman. There were other women but their DNA didn't survive. Natural selection eliminates any mutations that are harmful because the individuals having them don't survive.

Finding Mitochondrial ...

Most of our DNA res... zed in 23 pairs of chromosomes... enome in long strings of more t... d into at least 70,000 genes (thi... rn more). Outside the nucleus is ... cells contains anywhere from 1... ntains its own small ring of DN... dria were independent bacterial ... , is made up of a mere 16,500 b... ion of the mitochondrion. Each ... icating its small genome exactly—...

The human egg ar... st like all of our other cells. Dur... st enough power to reach the sur... that combine with the 23 female chromosomes in the egg, but it otherwise does not enter the egg. The fertilized egg that becomes a new human contains only the female's mitochondrial DNA. Because we inherit only our mother's mtDNA, it is passed along only through the female lineage. Furthermore, mtDNA is passed along very cleanly with almost perfect duplication, generation after generation. This is very different from the shuffling and recombining that nuclear DNA experiences in the process of sexual reproduction.

However mtDNA does change slowly over many generations because of random mutations. In the simplest case, a single letter (a nucleotide) in the genetic code gets switched to another letter. This is called a *single nucleotide polymorphism* (or *SNP*). Most mutations like this are harmless, or neutral, but in rare instances one may be beneficial; natural selection eliminates any that are truly harmful because the individuals having them do not survive. If a woman acquires a specific neutral mutation in her mtDNA, she will pass this on to her children, and her daughters will pass it on to their children. This single mutation will mark her descendents as distinct from all other humans. Geneticists call such a mutation a *genetic marker*, and they can use these markers to group people in lineages that originated from a single mother.

Allan Wilson and his team at Berkeley in 1986 were the first to compare specific stretches of mtDNA in people from around the world, and they found that different groups shared certain markers. Europeans had specific markers that no one else did, Asians had unique markers, and Africans had the most diverse set of markers of all. From these comparisons, they were able to construct an evolutionary tree for humans, and at the trunk of the tree were Africans. They were also able to make time estimates by using the molecular clock technique, which Wilson had pioneered in the late 1960s. This assumes that mtDNA mutates at a known constant rate.* When the mtDNA of two individuals is compared, the number of differences is a measure of the time since the two had a common female ancestor. When Wilson's team did this for their sample of 147 humans from around the world, they found that all of them shared a common female ancestor who lived about 200,000 years ago in Africa; she soon came to be known as mitochondrial Eve.

Wilson's discovery was immediately misunderstood and criticized. Many people thought that he had found the first woman or even the Biblical Eve, and some scientists questioned the validity of the team's conclusions because only a small number of single mutations (SNPs) were used for comparison, and the Africans he sampled were mostly African Americans. Wilson was not, however, saying that this Eve was the first woman. In fact many other women were alive at the same time she lived, but none of their mtDNA survived. The descendants of these other women eventually had only sons, or no children at all; or some of these other lineages may have been eliminated when large groups of people died during the massive droughts that swept through Africa. Yet Eve's mitochondrial DNA did survive, and it resides in every one of us today.

Since 1987, others have duplicated and extended Wilson's work using better and faster DNA sequencing techniques, and his original conclusions have largely been supported. A study published in 2000 (Ingman et al.) compared the entire mitochondrial genome of 53 different people from around the world and concluded that mitochondrial Eve lived in Africa about 170,000 years ago, plus or minus 50,000 years—that is, somewhere between 220,000 and 120,000 years ago. Many researchers today accept a rough date of about 150,000 years ago for the time of *the most recent common matrilineal ancestor of all living humans*, which is what mitochondrial Eve really is.

* The mutation rate of human mitochondrial DNA has been the subject of considerable research and debate. In 2000 it was estimated to be about one mutation per 232 generations, or about one mutation every 4500 years. See Donnelly, Peter et al., The mutation rate in the human mtDNA control region, *The American Journal of Human Genetics*, 1 May 2000. However, in 2005 Mark Stoneking and Brigitte Pakendorf reported that different regions of the mtDNA ring had substantially different mutation rates. See Mitochondrial DNA and human evolution, *Annual Review of Genomics and Human Genetics*, 2005. 6:165-183.

Both the fossil record and the genetic record indicate that the last of the hominids, *Homo sapiens*, originated in Africa about 200,000 years ago. Yet the details of the next

195,000 years are puzzling and contentious, until about 5000 years ago when systems of writing emerged and the written history of humanity could begin.

Like their ancestors *erectus* and *heidelbergensis*, the first people (for we can finally call them that) moved around as hunter-gatherers, constantly in search of new food supplies. Between 200,000 and 100,000 years ago, people spread throughout much of Africa, from Cape Horn to the Mediterranean and even to the Atlantic coast in some places, but apparently in small isolated groups. Survival was tenuous for these early humans, and their population remained small because of severe drought conditions that lasted from about 135,000 to about 75,000 years ago. This megadrought shows up clearly in core samples taken from the bottom of Lake Malawi in Southwestern Africa (Scholz et al. 2007). Lake Malawi, which today is about 600 meters (2000 feet) deep, lost 95% of its water at times during the megadrought and humans must have been pushed repeatedly to the brink of extinction. Not until about 70,000 years ago did the climate become wetter again so that the human population could grow.

Paleoanthropologists have very limited knowledge about what early humans were doing between 200,000 and 100,000 years ago because very little physical evidence has been found from this period. There is, however, a growing belief that humans made a significant developmental change during this period, perhaps beginning around 150,000 years ago. In 2007 an international team of researchers published their findings from caves that were occupied by early humans at Pinnacle Point on the southern coast of Africa (Marean 2007). Here, some 125,000 years ago, people began eating food from the sea, using pigments for symbolic purposes, and making the first bladelet tools. By about 100,000 years ago, decorative seashells and ochre pigments were being used as jewelry and symbolic art in locations as widely separated as South Africa, North Africa, and Israel (Vanhaeren et al. 2006). These are all signs that culture was evolving and symbolic thought was beginning to emerge in humans.

Adding to the physical evidence that paleoanthropologists dig up in the trash piles and burial mounds of early humans is the genetic evidence that is now emerging as molecular geneticists analyze the DNA of living humans. In the spring of 2008, a team of researchers from around the world published the results of a large-scale analysis of the complete mitochondrial genomes of over 600 living Africans (Behar et al. 2008). They found that the early human population had split into two main groups by about 150,000 years ago, one living in South Africa and the other living in Eastern Africa (present day Ethiopia). There is also evidence of many other small isolated groups throughout Africa who did not survive. All humans living today are descended from the two early populations, with most of us coming from the East African side of the split. The Khoisan Bushmen of South Africa are the only survivors of the South African side of the split (Figure 10-1).

Figure 10-1

A detailed evolutionary tree of humans for the last 200,000 years, derived from mtDNA studies published in 2008(Behar et al.). The final migration out of Africa that gave rise to all modern non-Africans is shaded in maroon. (LSA=Late Stone Age)

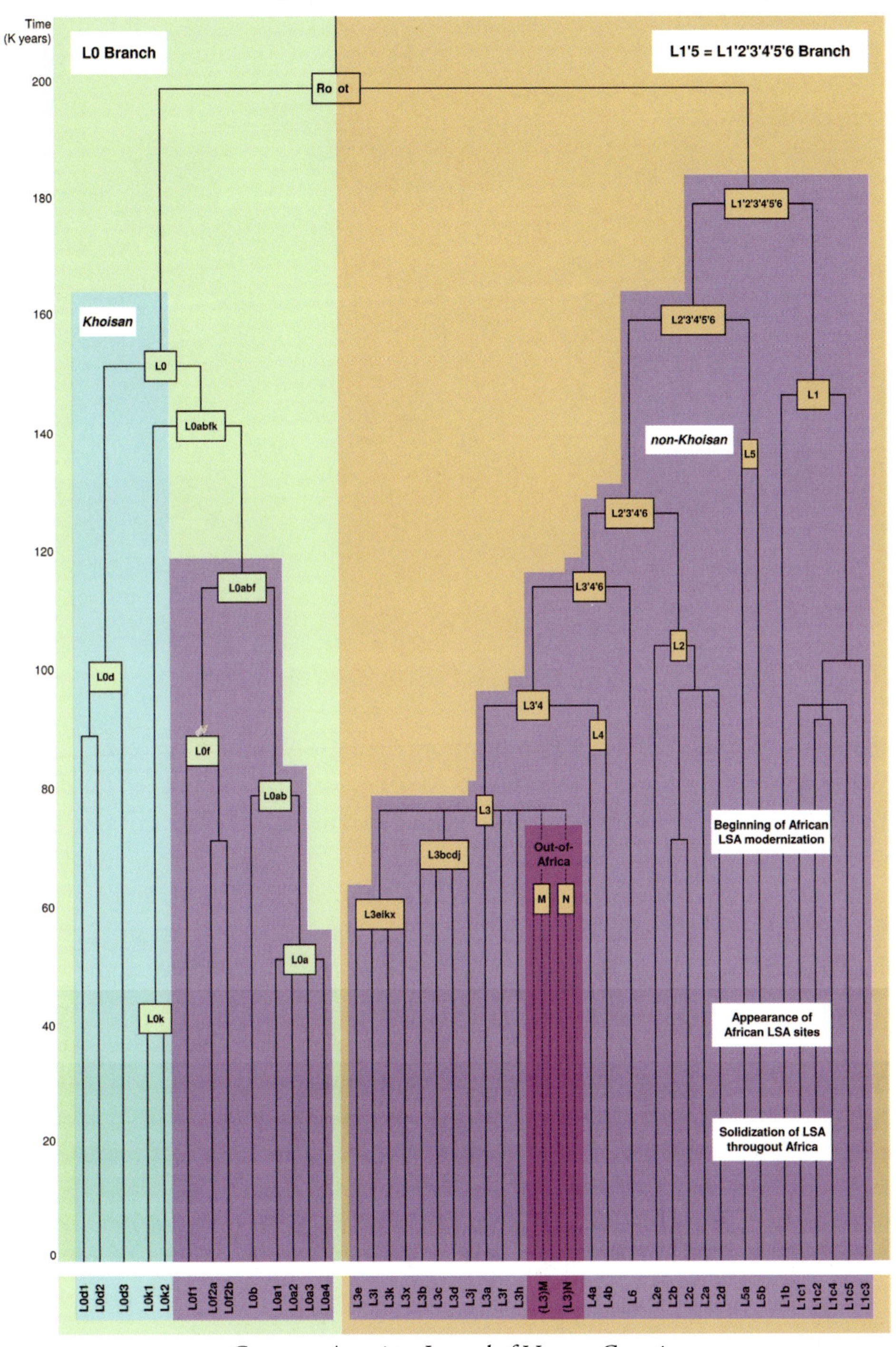

Courtesy: American Journal of Human Genetics

Figure 10-2

Two theories of early human migrations within Africa, before the final exodus about 60,000 years ago. From mtDNA studies published in 2008 (Behar et al.).

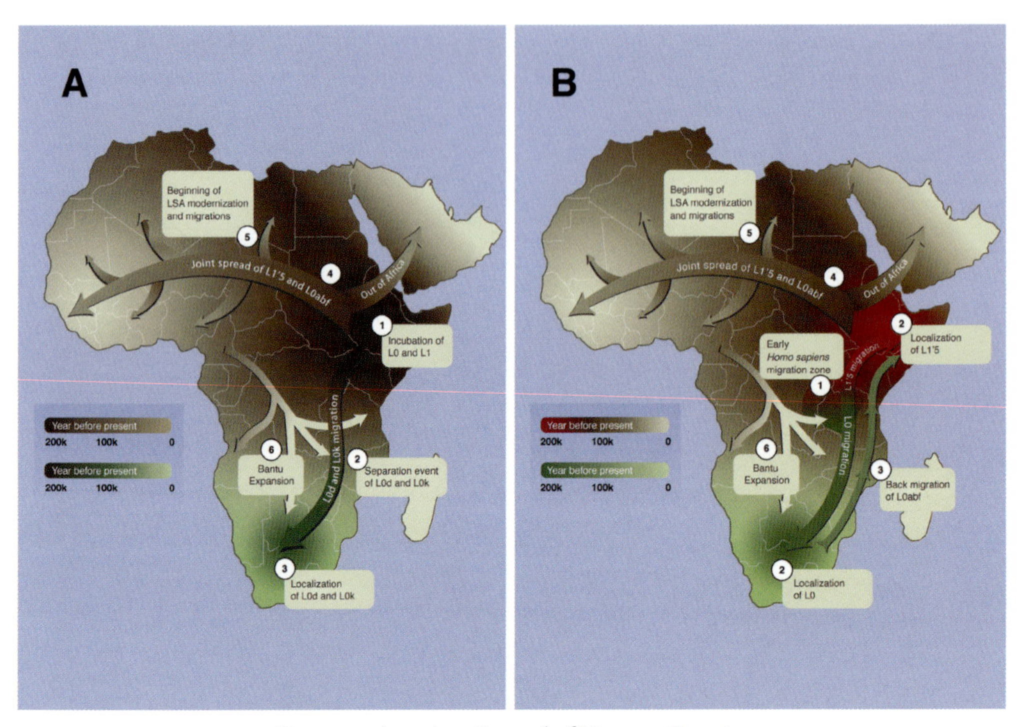

Courtesy: American Journal of Human Genetics

By about 100,000 years ago, humans from the East African group living in modern-day Kenya and Ethiopia began to leave Africa, perhaps in response to the deteriorating environment. Two different groups, using two widely separate exit routes were apparently successful in this bold and difficult exodus. A northern group followed the Nile River to the Mediterranean Sea, then crossed the Sinai Peninsula and moved into the Levant (a historical name for the region around modern day Israel and Lebanon). Human remains dating to about 92,000 years ago have been found at the Qafzeh Caves in Israel, leaving no question that early humans got that far out of Africa.

Apparently that was as far north as this group went, because there is no evidence of humans north of the Levant from this time. Something stopped them, and that something may well have been the Neanderthals, whose territory they were moving into. There is also clear evidence of Neanderthal presence in the Levant at this time. The human population was probably small and struggling for survival, while the Neanderthals were physically stronger, had a comparable level of tool technology, and a larger, well-established population that covered the Middle East, Western Asia, and Europe.

Some scientists have speculated that there was interbreeding between the two populations, based on skeletal remains that have a mix of Neanderthal and human characteristics. Until recently, the genetic evidence from comparing the human and

Neanderthal genomes did not support this possibility, instead suggesting that there was no mixing of genes, that the two were distinctly separate species incapable of producing viable offspring. However in 2010 this view was reversed when genome studies of living humans showed that people of Eurasian descent carry Neanderthal genes—anywhere from 1 to 4 percent (Green et al. 2010). It now seems that humans and Neanderthals must have interbred somewhere along the way, perhaps during this first encounter in the Levant. No one yet knows what the relationship was between Neanderthals and humans, but what does seem clear is that this first group of humans in the Levant did not (or could not) move any further north. In fact they may not have survived at all in the end, as we will see.

By about 100,000 years ago, another group of humans left Africa by a different route, a thousand miles south at the Strait of Bab-El-Mandeb on the Red Sea. Here the Horn of Africa nearly touches the Arabian Peninsula, separated by only 17 miles of water. There is uncertainty about how far this southern group moved into the Arabian Peninsula, but there is evidence that some of them made it all the way into India and perhaps China,* probably following coastlines and living as "beach combers."

It is clear that by about 75,000 years ago, humans had successfully moved out of Africa and should have been ready to populate the Earth. Then something went terribly wrong for *Homo sapiens*, because the genetic evidence suggests that the entire human population was very nearly wiped out, and only a small number of humans would survive the next few thousand years; they were the people who had stayed back in Africa. Those few African survivors are the common ancestors of all of us today.

Bottleneck

When scientists in the early1990s began looking at the nuclear DNA from many living humans, it became evident that our ancestors must have gone through something that reduced our total population to a very small number—a *population bottleneck*. Because the genes of all humans living today are found to be 99.9% similar, it must be that all of us are very closely related and had a common ancestor not long ago. By comparison, there is more variation in the genes of one social group of 50 chimps than in the entire human population.

---

* Michael Petraglia and his team found stone tools below and above the layer of volcanic ash from the eruption of Mount Toba (73,000 years ago) at a site in Jwalapurim, India, suggesting that the users of these tools were in India before 73,000 years ago, and that they survived the Toba event. The users are presumed to be humans.

See Petraglia, Michael, et al., Middle Paleolithic assemblages from the Indian subcontinent before and after the Toba super-eruption, *Science*, 5 July 2007.

Also, in 1958 a human skull was found in Liujiang, China and tentatively dated to 20,000-30,000 years ago, but more recent dating of the presumed site (there is some uncertainty about this) has produced a range of ages from 68,000 to 150,000 years old, long before humans were thought to have been in China. See Bower, Bruce. 2002, Chinese roots: Skull may complicate human-origins debate, *Science News*, 21 December.

Y chromosome studies on living males from worldwide populations now suggest that modern humans did not successfully leave Africa until about 60,000 years ago (Stix 2008) (see *Exploring Deeper: Y Chromosome Adam, p. 175*). This means that none of those first ventures out of Africa between 100,000 and 75,000 ago years were successful in the long term because none of that genetic material survived. It seems that the entire human population was culled back to just a few thousand members who lived in Eastern Africa and became the founding population of all of modern humanity. Our species was on the brink of extinction about 70,000 years ago.

What could have caused this population bottleneck that nearly wiped out humanity? Two factors are now thought to have contributed to our near-extinction: a mega-drought that lasted from 135,000 to 75,000 years ago probably created a series of bottlenecks that kept the human population small and scattered; and something even more deadly was first suggested by Stanley Ambrose from the University of Illinois in 1998: the eruption of Mount Toba on the Island of Sumatra about 73,000 years ago (Ambrose 1998).

The Toba event shows up clearly as thick layers of volcanic ash across India, and in ice cores from Greenland with elevated levels of atmospheric sulfuric acid. This was the largest volcanic eruption of the last 28 million years, spewing out lava that flowed for hundreds of miles and hurling gases, dust particles, and ash into the atmosphere. Mount Toba ejected *2,800 cubic kilometers* of material from inside the Earth, compared to the Mount Krakatoa eruption of 1883 which produced only *ten cubic kilometers* of material, and Mount St. Helens in 1980 which produced a mere *one cubic kilometer*.

The super-eruption of Toba probably lasted a few weeks and produced a thick blanket of volcanic ash as much as thirty feet deep in some locations thousands of miles away. The material that was immediately hurled into the atmosphere did not spread out in all directions but was carried by the prevailing winds for thousands of miles to the west and north (see 10-3). This "kill zone" extended directly into nearly all the areas where early humans had migrated, including the Levant. Eventually the entire atmosphere of the Earth filled with dust, blocking the Sun and triggering a devastating volcanic winter. It is not known whether the people who were downstream from Toba survived the initial effects of the eruption, but any that did survive would have faced a volcanic winter of extreme cold for the next five to ten years, and ice age conditions that brought widespread drought and famine for the next thousand years. Plant samples from the Levant at this time show a drastic change from an Afro-Asian community to a paleoarctic biome, typical of a much colder climate. For people who were hunter-gatherers with no reserves of food, survival would have been nearly impossible. The effects of Toba lingered for thousands of years before Earth's climate finally began to warm and the human population recovered.

At the time of the Toba eruption 73,000 years ago, there were at least three species of hominids living on the Earth (Figure 10-3). The Neanderthals were well established from Europe to Central Asia and south to the Levant. *Homo erectus* occupied Asia, from China to New Guinea. And the newcomers—*Homo sapiens*—lived in Africa and some had spread north into the Levant, east into the Arabian Peninsula, and perhaps into India and China. Two of these three hominid species—the Neanderthals and *erectus*—were spared the brunt of Toba's eruption because their territory was largely out of the kill zone. The volcanic winter and ice age that followed probably forced them closer to the equator—the *erectus* population moved toward Indonesia and New Guinea, while the Neanderthals moved into southern Europe and the Levant. The Neanderthals were already cold-adapted and probably handled the effects of Toba better than their cousins.

Figure 10-3

The scene of Homo sapiens exploits between 195,000 years ago and 73,000 years ago when Mt. Toba erupted. See more detailed key below.

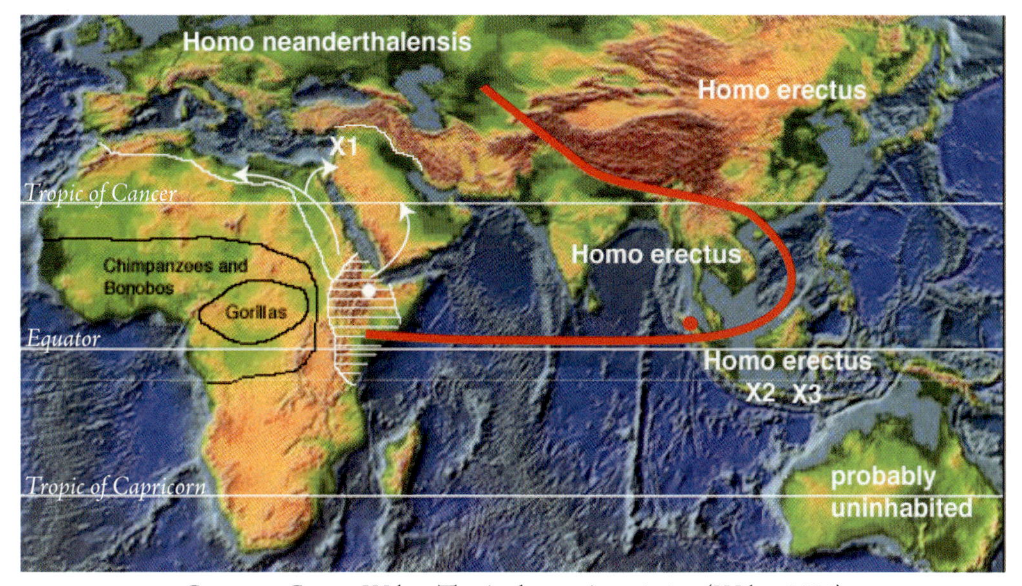

Courtesy: George Weber, The Andaman Association (Weber 2004)

<u>Red Dot:</u> Mount Toba

<u>Red Line:</u> The Toba kill zone where ashfall may have had a lethal and almost instant impact. In India the thickness of the ash layer found today ranges from 10 feet to 20 feet, and in Malaysia as much as 30 feet. Further west, the ash-falls would have thinned gradually, but dust and aerosol clouds would still have blocked the sun and brought on a calamitous drop in temperature—the forerunner of volcanic winter.

<u>White Dot:</u> Herto, Ethiopia, site of one of the oldest known Homo sapiens idaltu, dating from 154.000 to 160,000 years ago. Close by is Kibish, Ethiopia with human remains dating to 195,000 years ago.

<u>White shaded area:</u> Area occupied by the first anatomically modern Humans. 200,000-150,000 years ago

<u>White-bordered area without hatching:</u> Vulnerable areas into which Homo sapiens had expanded by about 100,000 years ago until the Toba eruption. The limits of this expansion are highly uncertain and include only the Levant for sure. Whether Mesopotamia, the Arabian Peninsula, and India were also settled is not yet certain but it remains a possibility.

<u>White X1</u>: The Levant, site of Homo sapiens/Homo neanderthalensis interaction about 95,000 years ago.

<u>White X2</u>: Toba survivor Homo erectus on Java, Indonesia

<u>White X3</u>: Toba survivor Homo floresiensis on Flores, Indonesia

The humans, however, who had moved out of Africa by the time Toba errupted had placed themselves directly in the kill zone (10-3, previous page). If any survived the initial dump of ash, the extreme cold that followed may have finished them off, or perhaps forced some of them back into Africa. The humans who had remained in Africa, especially southern Africa, were much safer from the immediate effects of Toba but still had to struggle to survive the extreme drought. It seems that a small group of Africans, perhaps only a few thousand, became the founding population of modern humanity, and the common ancestors of all humans living today. The combination of the African mega-drought and the Toba event must have been the final bottleneck that humans had to pass through—a critical test of survival—before spreading over the entire Earth.

By about 70,000 years ago the mega-drought was over, the global winter caused by Toba was subsiding, and the human population in Africa began to recover. Those who had made it through the bottleneck had powerful survival advantages, probably involving higher intelligences, and as survivors they were able to pass along these advantages to their offspring. A sophisticated social intelligence and the mastery of spoken language may have allowed people to work together cooperatively and share knowledge. Ambrose has proposed that a "troop to tribe" transition was solidified by the Toba event that marked the beginning of complex social networks and an explosion of human culture. The survivors developed many new technologies like clothing, shelters, and better hunting tools, all signs of a higher intelligence.

Although the Toba event very nearly caused the extinction of our species, it may also have catalyzed a reorganization of the brain that brought an explosion of higher intelligence. There are multiple lines of evidence showing that by about 60,000 years ago humans were behaving in significantly different ways. The African hominid that came through the Toba bottleneck was unlike anything that had lived before.

The Great Leap

In the 1992 book *The Third Chimpanzee*, Jared Diamond called this major change in human behavior "the great leap forward" and dated it to about 40,000 years ago when elaborate cave art and a new generation of sophisticated tools began to appear in Europe. Richard Klein of Stanford University later called it "the big bang of human consciousness" in his 2002 book *The Dawn of Human Culture*, and dated it to about 50,000 years ago, based on more recent evidence from Africa. According to Klein, "Humanity was transformed from a relatively rare and insignificant large mammal to something more like a geologic force"(Kline and Edgar 2002). More recent studies suggest that this momentous change was not so sudden, as the terms "leap" and "big bang" connote, but came about gradually over many tens of thousands of years, reaching fruition by about 60,000 years ago.

The great leap is clearly evident when we compare human artifacts in Africa from between 200,000 and 100,000 years ago to those that are about 60,000 years old. Paleoanthropologists call this earlier period the *Middle Stone Age* and the later period the *Late Stone Age*. The artifacts and behaviors of the Middle Stone Age people were very much like their Neanderthal cousins to the north. The stone tools of these Africans are called *Mousterian* tools and are also found in Europe where the Neanderthals lived. These tools were made from core stones by carefully striking off flakes with another stone. Between 200,000 to 100,000 years ago, there was a small range of tool types that hardly changed over this time, and tools were very similar over many locations. The stone they used was from local sources, which indicates that they had a small range of movement and no long distance trade; they did not use bone, ivory, or shells; these Middle Stone Age people buried their dead, but without ceremonial grave goods; there is no clear evidence of structures in their campsites, and they were fairly ineffective hunters who could kill only docile animals and not dangerous animals or birds; there is also little or no evidence of fishing technologies.

By about 60,000 years ago, human tools in Africa had changed drastically. These Late Stone Age people made a wide variety of tools including needles, fishhooks, borers, and knives. Along with stone, they also used bone, ivory, and shells. Tools varied from one location to another and changed noticeably over time, showing a much higher level of innovation. Art objects like figurines, pendants, and beads are common for the first time; their burials included lavish grave goods like jewelry and beads, suggesting elaborate ritual; solidly built houses, tailored clothing, and more efficient fireplaces are found; their much better hunting technology and skill allowed them to kill buffalo, wild pigs, birds, and fish. While the simple stone tools of the Middle Stone Age hardly changed between 200,000 and 100,000 years ago, the sophisticated Late Stone Age technology improved rapidly, reflecting a significant leap in human development.

However, the size and shape of the human brain had not changed appreciably for hundreds of thousands of years. Something else must have changed. What was it, and how did it happen? Richard Klein believes that the only explanation is a "fortuitous mutation that promoted the fully modern brain." He proposes that this mutation brought the capacity for symbolic thought and rapidly spoken language. Some kind of reorganization or rewiring must have occurred in the brain, and it brought a social intelligence to humans based on symbolic thought and language. The rapid development of new technology was something that no other living thing had ever achieved.

Others argue that this was not so much a genetic change, or mutation, as an *evolution of culture*. Peter Richerson and Robert Boyd make the case in their book *Not by Genes Alone* that human culture was evolving because people could learn from each other and pass along customs and knowledge—such as how to make and use tools—from generation to generation, and build on that knowledge. Natural selection favored this evolution of culture because it gave humans very powerful survival advantages. Richerson and Boyd define culture as " information capable of affecting individuals' behavior that they acquire from other members of their species through teaching, imitation, and other forms of social transmission" (Richerson and Boyd 2006).

More simply, culture is that which we learn from others. This is what modern humans could do better than any other living thing: learn from each other, build upon previous knowledge, and teach new knowledge to others. The ever-increasing human knowledge base transcends any one individual or lifetime, because it can be passed along to new generations in much the same way we can pass along our genes to a new generation. The big hominid brain seems to have been sitting inside the skull, largely unused for hundreds of thousands of years, but now with this explosion of culture, with the emergence of symbolic thought and language, and with the acquisition of knowledge, the brain began to function more fully. We can only wonder today how much more the brain is capable of, and how much further human culture and consciousness can evolve. There is no reason to think we have "arrived" at our full potential.

One of the best lines of evidence for the emergence of symbolic thought is the ostrich shell beads that have been found from Kenya to South Africa, some as old as 75,000 years. These are difficult and time-consuming to make, but are found so widely that they must have had some special significance. Stanley Ambrose, who discovered ostrich shell beads in Kenya in the 1980s, has suggested that they were highly important gifts that were exchanged by neighboring tribes of people in an act of reciprocity. Should food be in scarce supply for one group, they could move into another group's territory if reciprocal ties had been established. "They're paying into their health insurance, in a sense," says Ambrose (1998). "They're paying insurance to each other. With this social safety net they could do better than people without symbolic means of establishing future permanent ties of reciprocity." The beads symbolized good relations with other people and may have provided the confidence to move into riskier environments.

According to Klein, "Once symbols appear in the archeological record, as enigmatic geometrical designs, as human or animal figurines carved in ivory, or as beads and other ornaments, we know we are dealing with people like us: people with advanced cognitive skills who could not only invent sophisticated tools and weapons and develop complex social networks for mutual security, but could also marvel at the intricacies of nature and their place in it; people who were self-aware" (Klein and Edgar 2002).

Still no one knows exactly what caused this leap in intelligence and culture that marked a watershed change in humanity, only that it may have started by 160,000 years ago and had apparently culminated by about 60,000 years ago. Philosophers, theologians, and scientists have discussed and debated for centuries the qualities that make humans different from all other animals. Today most scientists agree that symbolic thought, speech, and language are key defining characteristics of modern humans. Speech and language reflect symbolic thought: when we write or say the word "tree," it symbolizes a real object. Speech brought about complex social interactions and the ability to accumulate knowledge. Until about 5000 years ago, knowledge could be shared only through speaking, gesturing, or imitation. Elaborate oral traditions of storytelling and performance emerged within tribes of people. The vivid mythical accounts of creation, heroism, and tragedy were passed along from generation to generation, bringing structure and meaning to the world. This mythical era of storytelling probably lasted from about 60,000 years ago until about 5,000 years ago, when *written* language emerge, and a whole new capacity for communication became possible.

The new human that emerged by about 60,000 years ago brought a whole suite of abilities to the game of survival that had never existed before. They could now do much more than just survive.

Figure 10-4
Some of the Unique Markers of Modern Humans after The Great Leap
(Some of these were acquired much more recently)

- Symbolic thought
- Spoken language
- The sharing and accumulation of knowledge
- Rapidly evolving culture
- Self-awareness* (an emerging ego)
- Creativity and the development of art and music
- Innovation and the rapid development of new technologies
- An emerging sense of time—past, present, and future
- Awareness of one's own life and death, and the need to know where we came from
- Curiosity and the ability to pose questions
- Sophisticated problem-solving ability
- Morality, a sense of good and evil, and the capability to do both
- The acquisition of power, first over nature, then over other people (about 5000 years ago)
- The concept of God or gods, addressed by worship, prayer, or denial
- Written language (not until about 5000 years ago)
- The development of permanent fields of knowledge, like mathematics, medicine, and astronomy (not until 4000 years ago)
- The development of empirical science (not until 400 years ago)

---

* Chimpanzees have shown a limited degree of self-awareness and morality.

Chapter Eleven
Sapiens: Inheriting the Earth

Out of Africa (again): The Peopling of the Earth

Modern humans have been living on Earth for about 200,000 years, and for most of that time we lived only in Africa. The genetic evidence now suggests convincingly that all humans living on Earth today are descended from a small group of Africans who lived about 60,000 years ago in what is now Ethiopia. This can only mean that the earlier humans who got out of Africa about 100,000 years ago left no descendants who are alive today. We may never know the whole story of why that is; perhaps they perished in the volcanic winter of Toba, or maybe the later modern humans replaced them as they spread over the Earth.

The human that emerged from the Toba bottleneck by about 60,000 years ago, *Homo sapiens sapiens*, had a suite of abilities never before seen on Earth. They were like us in many ways, both in appearance and in intelligence. While the early modern humans in Africa barely survived from 200,000 years ago until about 60,000 years ago, now nothing could stop them. As their population began to grow, they once again left Africa, probably by the more southern route, and moved into the Arabian Peninsula (Figure 11-1). They apparently did not move directly north to the Levant and into Europe at this time, but continued to move east into Asia where they showed up in Indonesia and China by 55,000 years ago and Australia by perhaps 50,000 years ago. From here, some =moved north into Central Asia by 45,000 years ago, and then west into Europe by 40,000 years ago. The final diaspora that spread humanity all over Asia, Europe, and eventually into the New World was relatively rapid. Nothing could stop the expansion of these new humans as they gained and shared knowledge and developed increasingly powerful technologies.

Exploring Deeper

Y Chromosome Adam

Just as mitochondrial DNA can be used to trace female lineages (see *Science and Discover: Finding Mitocondrial Eve*, pg. 166), males have a unique set of genes in the Y chromosome that can be used to uncover their story. The Y chromosome is the male half of the 23rd pair of human chromosomes, and is carried by the sperm and passed along only from father to son. Like all DNA, it also mutates over time as it is passed down through many generations. However, the Y chromosome is much

Firure 11-1:
Out of Africa: The peopling of the Earth.

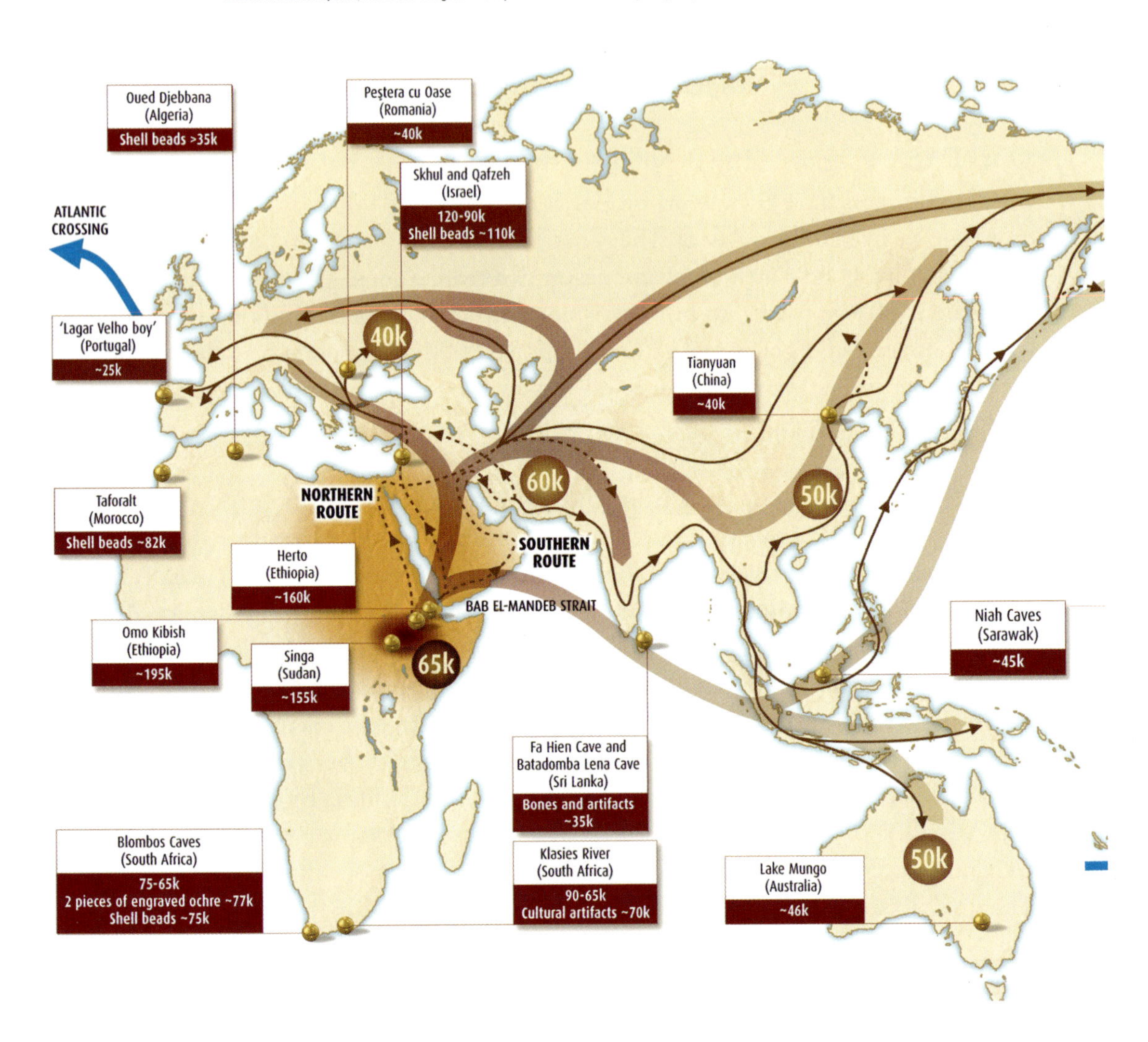

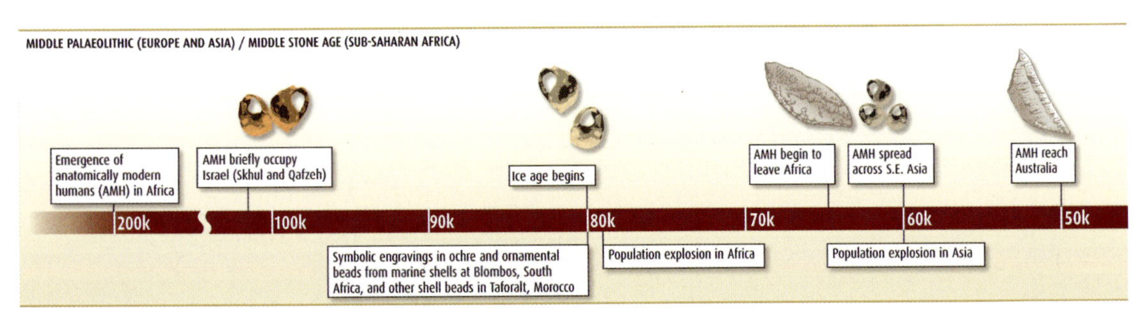

BERINGA ROUTE
16k
PACIFIC COASTAL ROUTE
PACIFIC COASTAL ROUTE
Wally's Beach (Canada)
Human occupation ~13k
ATLANTIC CROSSING
Arlington Springs (US)
Leg bone ~13k
Cactus Hill (US)
Small blades 20-17k
South-east of Puebla (Mexico)
Ancient footprints ~38k (contested)
Flow of genes around globe
Routes of migration
Alternative/contested routes
~13k Anatomically modern humans
10k 10,000 years ago
Quebrada Jaguay (Peru)
Human settlement ~13k
©NewScientist
PACIFIC CROSSING
15k
Monte Verde (Chile)
Flake tools and stone points ~15k

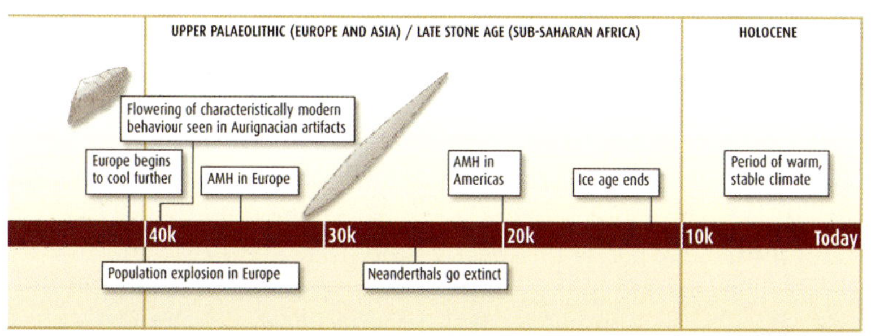
UPPER PALAEOLITHIC (EUROPE AND ASIA) / LATE STONE AGE (SUB-SAHARAN AFRICA)
HOLOCENE
Flowering of characteristically modern behaviour seen in Aurignacian artifacts
Europe begins to cool further
AMH in Europe
AMH in Americas
Ice age ends
Period of warm, stable climate
40k
30k
20k
10k
Today
Population explosion in Europe
Neanderthals go extinct

harder to work with than mtDNA because it is made of 60 million base pairs rather than 16,500. In 1995, researchers at Stanford pioneered new and faster ways to read the Y genome (Cavalli et al. 1996). They could quickly look at long stretches of DNA as well as single nucleotide markers and use these to trace common ancestry. Like Allan Wilson before, they constructed an evolutionary tree for humans and found once again that the trunk of the tree was exclusively African. By the year 2000, researchers concluded that all living males were descended from one man who lived in Africa about 60,000 years ago—the so-called "Y chromosome Adam." By 2011, better analysis techniques led to a revised date closer to 140,000 years ago (Cruciani et al. 2011); and like mitochondrial Eve, Y chromosome Adam was not the first man, just the most recent common *patrilineal* ancestor of all males alive today.

Y chromosome studies of living males from different parts of the world have also revealed the ancient migration routes of humans out of Africa, starting about 60,000 years ago (see 11-1, previous page). This picture is largely in agreement with the physical evidence of human migrations.

The conclusions reached from recent genome studies remain controversial among some paleoanthropologists. Concerns have been raised about mutation rates and how constant they really are, about contamination of ancient DNA samples with modern DNA from handlers, about the possible mixing of mitochondrial DNA with nuclear DNA, and possible environmental selection on genes (for example when a group of people migrates from a hot climate to a cold one). Analysis of DNA is still an extremely difficult and complex undertaking, and certainly the field of molecular genetics is still in its infancy today. Results will become more reliable as techniques and technologies continue to improve. According to Mark Stoneking, describing the state of the art in 2005:

> It has therefore become abundantly clear that studies of mtDNA variation need to be complemented with data on the male-specific Y chromosome, and ideally with autosomal* data as well. (Pakendorf and Stoneking 2005)

The first modern people in Europe date to about 40,000 years ago and are often called *Cro-Magnons*, after the site in France where they were first discovered in 1868. Cro-Magnon culture in Europe is well documented, unlike human exploits of the same period in other parts of the world. The limestone caves that are common throughout Europe were ideal for preserving ancient artifacts and remains, and Europe has been explored more thoroughly than other parts of the world.

The Cro-Magnons had a wide array of behaviors and abilities that were new, including the use of advanced weaponry, the formation of long-distance trade networks, self-expression through art and music, and elaborate burials. The lavish cave paintings at Altamira in Spain, and Lascaux and Chauvet in France, are remarkable for their impressionistic style and innovative use of perspective and natural relief (Figure 11-2).

---

* Autosomes are all the chromosomes *other* than the sex chromosomes. For humans, this is the first 22 chromosomes, since the 23rd pair is the sex chromosomes.

Figure 11-2

Cro-Magnon Cave Paintings from 14,000 to 25,000 Years Ago

(a) Replica of painting from Altamira Cave, Spain dating from about 15,000 years ago. The cave was discovered in 1879.

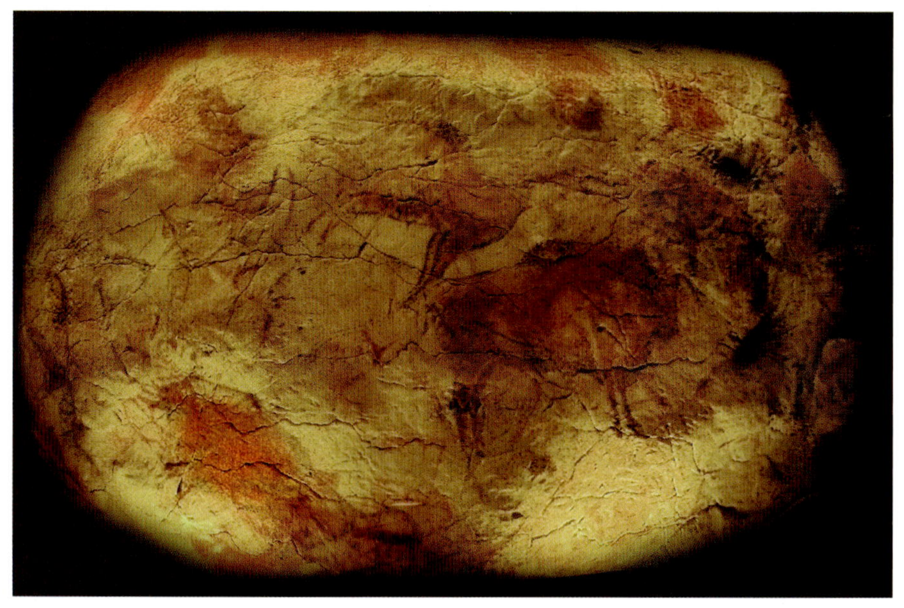

Wikimedia Commons/ National Museum of Archaeology, Madrid, Spain.

(b) Painting from Lascaux Cave, France, dating from about 14,000 years ago and discovered in 1940.

Wikimedia Commons

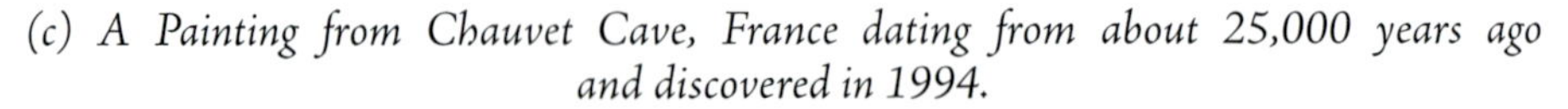

(c) A Painting from Chauvet Cave, France dating from about 25,000 years ago and discovered in 1994.

Wikimedia Commons

The Neanderthals were already well established when Cro-Magnon humans arrived in Europe about 40,000 years ago. In the first round of human-Neanderthal interactions in the Levant about 100,000 years ago, during the first migrations of people out of Africa, the early humans apparently could not out-compete or overpower the Neanderthals. They could not spread beyond the Levant and apparently did not even survive there. It was a different story, however, for the newly rewired humans who left Africa about 60,000 years ago. This behaviorally modern human, this new human, had undergone a leap in intelligence and culture, and once they came face to face with the Neanderthals in Central Asia about 50,000 years ago, the Neanderthals gradually began to disappear. Modern humans and Neanderthals coexisted for more than 20,000 years in Europe and Central Asia, but no one knows much about their relationship. Early on, their tools were very similar, so they must have shared knowledge to some extent. There does not seem to be evidence of slaughter, and recent comparisons of the Neanderthal genome with modern human genomes show that there was some interbreeding somewhere along the way. Svante Pääbo's team found that people of European descent inherited between 1% and 4% of their DNA from Neanderthals (Pääbo et al. 2010).

Cro-Magnon humans had probably developed extensive social networks and sophisticated speech, and may simply have viewed the Neanderthals as primitives capable of little more than grunting (although there is growing evidence that

Neanderthals may have been capable of speech to some degree*). The Cro-Magnons were probably just superior intellectually, socially, and culturally, and must simply have out-performed the Neanderthals as they competed for food and shelter. As the Cro-Magnons proliferated, the Neanderthals retreated further west and south in Europe, making their last stand in modern Spain and Portugal. Physical evidence of Neanderthals disappears altogether by about 28,000 years ago. Perhaps they were outcompeted into extinction or perhaps they were simply assimilated by the human population, or maybe both.

Our knowledge of human activities outside of Europe at this time is very limited, but there is no doubt that humans spread relatively quickly to every habitable continent. Human presence in Australia dates to as early as 50,000 years ago (this is still under debate), but how people got there remains a mystery. Most of Indonesia, New Guinea, and Australia were connected by land at this time because sea levels were low, but people would still have to cross about 50 miles of open water to reach Australia; it seems that they must have built boats of some kind, yet the first undisputed evidence of watercraft comes from the Mediterranean almost 40,000 years later. The crossing to Australia is remarkable considering that no land can be seen across 50 miles of ocean; so how did they know that there was land out there when they set out in their primitive boats, and not just thousands of miles of open ocean? Perhaps someone was accidentally carried across on a floating log and was able to return and tell others.

The First Americans

The peopling of Siberia was a complex and lengthy process that probably began some time between 45,000 and 40,000 years ago. Evidence from archeological sites, genome analyses of living Siberians, and linguistic studies show there were a series of migrations into Siberia from Southern Russia, Eastern Europe, Central Asia, and Mongolia. At the peak of the last ice age 22,000 years ago, people were living in northeastern Siberia in brutally frigid conditions. They stitched together warm, tailored hide garments using sinew and bone needles, and they built shelters, used fire, and possessed effective hunting weapons. Eventually some of them crossed what is now the Bering Strait and moved into Alaska, entering the last unsettled continent. With sea levels as much as 100 meters lower, the present Bering Sea was a land bridge.

Exactly when and how people moved into the New World is still uncertain. Until recently it was believed that humans would not have been able to move south from Alaska into modern day Canada and the northwestern U.S. until about 13,000 years ago because a massive ice sheet covered all of this area. It was presumed that the first

* In 2007 the FOXP2 gene, implicated in human speech, was found in the Neanderthal genome. See Krause, Johannes, et al. The derived FOXP2 variant of modern humans was shared by the Neanderthals. *Current Biology*, November 6, 2007.

people in the present day lower 48 states arrived through an inland passage between ice sheets about 13,000 years ago. These were the Clovis people, named after the distinctive spear points found in 1932 near Clovis, New Mexico. Hundreds of Clovis sites have been found with the signature spear points, or Clovis points, suggesting that the Clovis culture covered most of the United States with a population that grew to perhaps 10 million. However, new archeological sites have produced abundant evidence that this *Clovis First* theory is not correct.

It is now certain that pre-Clovis people lived in North and South America. Physical evidence of pre-Clovis Americans comes from Paisley Cave in Oregon and dates to 14,300 years ago, a thousand years before the Canadian ice sheets began to disappear (Gilbert et al. 2008). These early people must have either moved down the rugged coast if it was ice free, or perhaps they used watercraft to move south. Earlier still is evidence of human presence from 14,600 years ago in Monte Verde, Southern Chile, several thousand years before humans were thought to have been there (Dillehay et al. 2008). Finally putting the Clovis First theory to rest are the discoveries from a massive pre-Clovis site at Buttermilk Creek, Texas, with artifacts dating to 15,500 years ago (Pringle 2011).

Recent genetic evidence, complementing the physical evidence from these and other archeological sites, has begun to paint a more complete picture of how and when humans moved into the Americas. An international team led by Antonio Torroni of the University of Pavia in Italy analyzed genetic markers in the mitochondrial DNA of modern-day Native Americans, and concluded that humans had reached Alaska by about 16,000 years ago, and that by about 15,000 years ago, two separate groups had reached Wisconsin and the coast of Chile (Balter 2009). Other genetic studies (First Americans 2009; Perego et al. 2009) support the scenario of two separate groups of people crossing from Siberia into Alaska between 15,000 and 17,000 years ago and migrating southward into the Americas. One group traveled the Pacific coast all the way to Patagonia at the southern tip of South America and another group spread inland throughout North America to become the Clovis culture and all the later indigenous tribes of North America.

There have been alternate theories of how humans first got to the New World. One is that they came from Europe across thousands of miles of the frozen North Atlantic. Those who support this idea point out that the Clovis technology of North America resembles the *Solutrean* tool culture of France more than that of Siberia. It's conceivable that Europeans could have followed the edges of ice sheets across the North Atlantic—we know that Inuit people in Alaska (Eskimos) have adapted to such extremes. However, this theory has not found many supporters. The trail of physical evidence from Siberia to Alaska and into the Northwestern U.S. is far more compelling, and recent genetic studies of indigenous Americans show Asian ancestry, not European.

The people that traveled down the Pacific coast kept moving south for generation after generation, perhaps using boats (although no physical evidence of boats has yet been found). In about 1000 years they apparently moved all the way from Canada to Patagonia at the southern tip of South America, a distance of about 8,000 miles. This would require them to move an average of only 8 miles per year, something quite trivial for people who often covered that distance in a day. Why did they continue to move south? This would be understandable while they were in North America, north of the equator, where they would be following the sun into warmer climates. Then after crossing the equator in South America, moving south would take them into colder climates. What led them all the way to Monte Verde, Chile and eventually Patagonia? A few scientists have speculated that these first South Americans did not arrive from the north, but came across the Pacific in rafts, but as yet there is no evidence supporting this theory.

The arrival of humans in many parts of the world coincided with the extinction of large mammals. According to Jared Diamond, "About 15,000 years ago, the American West looked much as Africa's Serengeti plains do today, with herds of elephants and horses pursued by lions and cheetahs, and joined by members of such exotic species as camels and giant ground sloths" (Diamond 1999). Yet the fossil evidence shows that by about 12,000 years ago, when the Clovis culture was thriving, nearly all of these large mammals had disappeared. Some scientists have suggested that these extinctions were due to climate changes and not humans, but it's hard to ignore the fact that these large mammals had thrived for millions of years and then disappeared rather suddenly at about the time humans arrived. The many skeletons of large mammals that have been found with Clovis spear points lodged in their ribs leave no doubt that these people were very effective hunters.

A similar trend is seen in Australia, where roughly 60 species of large mammals and birds disappeared after the arrival of humans. Apparently this was not so much from over-hunting as from drastic changes in the vegetation. Early Australians set massive fires for some reason (perhaps to clear the land), and the landscape was changed from mixed forests to shrubs and grasses. The story was much the same on other continents. Starting about 50,000 years ago, humans had developed an unprecedented ability to destroy natural habitats. This brought more than the extinction of many plant and animal species. Jared Diamond (2005) shows in his book *Collapse* that many later human civilizations fell because they ravaged and depleted their own natural environment.

By about 12,000 years ago, the age of migration was drawing to a close. Starting from Africa some 60,000 years ago, humans had now spread over the Earth, like seeds scattered widely over every habitable area of the Earth. The last ice age was drawing to a close and the longest known warm spell was about to begin. The conditions were perfect for something new.

The Neolithic Revolution

From the time *Homo erectus* first left Africa almost 2 million years ago, our ancestors had been nomadic hunter-gatherers, constantly on the move in search of food sources. With the end of the last ice age, another way to produce food was emerging. It was based on a new relationship with nature. Rather than just preying on other species, and often hunting or gathering them to depletion, people began protecting certain species and encouraging reproduction to create a more reliable food supply. This was the beginning of the domestication of plants and animals.

Dogs were probably the first domesticated species long ago. They probably lingered near the camps of hunter-gatherers in search of food scraps, and they may have acted as watchdogs, alerting the people to the presence of other animals. The first domesticated plant species were seed plants like wheat, squash, peanuts, barley, and lentils. Once people started storing grain, they probably found cats to be very useful in controlling mice and rats. Soon they found that certain other animals were easily domesticated—sheep, goats, pigs, and later cattle. When humans select and protect the plants or animals that they find most desirable—by saving and planting seeds from a plant that produces the biggest fruits, or by breeding the largest animals—they are influencing the course of evolution. The domesticated species eventually begin to diverge from the wild species as human selection overrides natural selection; here was the first form of genetic engineering.

The domestication of plants and animals marked the beginning of what anthropologists call the *Neolithic Period,* or the "New Stone Age." This was the last 5000 years of the Stone Age that had begun with *Homo habilis* some 2.4 million years ago and ended about 5500 years ago when bronze replaced stone as the highest medium of technology. In the Neolithic period, a whole new direction for humanity emerged, away from the nomadic hunter-gatherer lifestyle that had lasted for millions of years and toward a sedentary lifestyle with permanent settlements and new food sources. It was the development of agriculture more than anything else that underpinned the Neolithic revolution, and eventually led to organized cities, stratified societies, powerful rulers, vast empires, and all else that has come to be called *civilization.*

The earliest signs that the hunter-gatherer lifestyle was changing come from archeological finds in the Levant, near modern-day Jordan (Bar-Yosef 1998). By about 15,000 years ago, as the last glacial maximum was ending and the climate was warming, people began harvesting wild cereal grasses like oats and barley in the fertile Jordan Valley. They were becoming semi-sedentary foragers, living out of base camps and small settlements that were near sources of grain. By about 14,000 years ago, the *Natufian* culture emerged in the Levant with tools and customs that were typical of all Neolithic peoples several thousand years later. They were probably the first farmers. However, these first steps toward agriculture and permanent settlements were interrupted about 12,800 years ago by a harsh cold, dry period known as *the Younger Dryas* that lasted for 1300 years. The sophisticated Natufian culture virtually disappeared as people returned to the more mobile hunter-gatherer lifestyle.

Figure 11-3

Late Stone Age technology: Neolithic tools from Northwest Africa. The modern human is holding two well-sharpened blade tools designed for cutting, chopping, and scraping.

Courtesy: Eric D. Miller Collection

The end of the Younger Dryas, about 11,500 years ago, was the very last of the ice age periods, and Earth's climate has been unusually warm ever since. With the warming climate, people once again began to settle down and harvest wild cereal grains. Sophisticated granaries have recently been found in Jordan dating to this time (Finlayson and Kuijt 2009), showing that people were living in permanent settlements and needed to store and protect large quantities of grain. Sometime between 11,500 and 10,000 years ago, people began to experiment with selecting and collecting seeds from the best plants and planting them in protected areas. The crops improved as they domesticated plants in this way. In these early experiments with the domestication of plants, simple tools like sticks and rocks were used to loosen soil and poke holes for seed planting. This kind of practice is referred to as *horticulture*, and is essentially gardening on small plots of land, planting and watering by hand. Because planting and tending crops demanded a continuous presence, permanent settlements and a more sedentary lifestyle were inevitable. Evidence of plant domestication comes from throughout the region of Eurasia known as the Fertile Crescent, an arc that extends from modern day Egypt to Iraq (see Figure 11-4). Although much of this area is arid and desert-like today, it was lush and green 11,000 years ago.

Figure 11-4
The Fertile Crescent

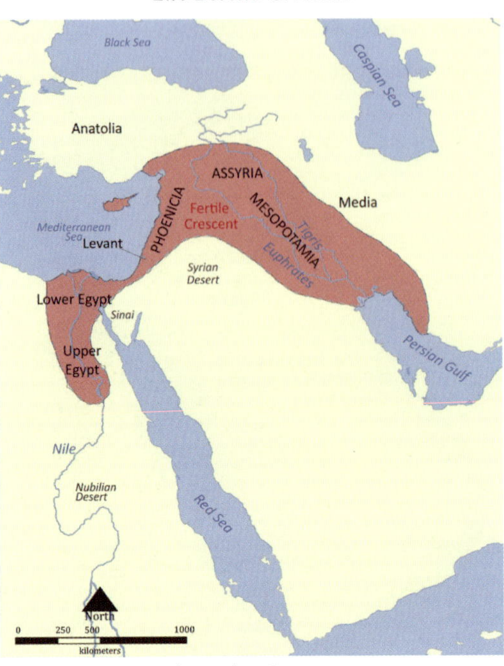

Wikimedia Commons

By about 11,000 years ago, when the first signs of domestication appear in Eurasia, humans were spread across the Earth in three separate world zones. The largest was the Afro-Eurasian zone where people moved around and shared ideas within a large area that included Africa, the Middle East, Europe, and Asia. The second world zone was made up of Australia, Indonesia, and New Guinea, and the third zone was the Americas, from Alaska to Patagonia. While there was a significant flow of ideas and goods *within* each of these zones as people moved about and traded, these three zones must have been completely isolated from each other. Yet, remarkably, the domestication of plants and the practice of horticulture occurred independently in all three zones, starting in Eurasia.

The earliest known large permanent settlement was Jericho, in the present day Palestinian Territories, and it was inhabited by about 10,000 years ago. Other large settlements like Catalhoyuk in Anatolia (modern Turkey) appeared by about 9500 years ago. At times Catalhoyuk contained as many as 10,000 people living in a honeycomb-like maze of houses with mud brick walls and horizontal timbers supporting a mud roof. The continuous flat rooftops became ground level where people walked. They entered a house by climbing down a ladder through a hole in the ceiling.

The structure of ancient Jericho was much the same, but Catalhoyuk was eventually abandoned, while Jericho has been continuously occupied for the last 10,000 years. These were not yet true cities, but disorganized collections of villages where both hunter-gatherers and farmers lived in close quarters. In these early *proto-cities,* there were no signs of social stratification, like wealth, rulers, or levels of status. The many dwellings were remarkably similar, and there were no temples or other public buildings.

The cultivation of crops became more efficient and productive as better technologies and techniques were used to clear, plow, and irrigate larger tracts of land. This was the beginning of *agriculture.* The food surpluses that came with large-scale agriculture meant that larger populations could be supported and that division of labor was possible for the first time. It was no longer necessary for every person to devote themselves to acquiring food—now people could specialize as builders, child tenders, healers, craftsmen, artists, farmers, or foragers. As populations grew, more people were available for work, and more types of work could be done.

From the time the *Homo* lineage began some 2.5 million years ago, our ancestors had always been tool users, and starting about 60,000 years ago the pace of innovation and discovery has been ever-increasing. It was in the proto-cities of Central Asia about 8000 years ago that one of the greatest of all technological discoveries was made: the use of metals. The first metal to be used was gold because it was fairly easy to mine and melted at low temperatures. However, it could be used only for decorative purposes, like jewelry, because it was so soft. Copper, a somewhat harder metal, was being used in Anatolia and Serbia by about 7000 years ago. With new techniques to make hotter fires, copper ore could be smelted into pure metal and, unlike gold, it was hard enough to be used for a wide variety of tools.

By about 5500 years ago, people in Mesopotamia discovered that adding tin to copper in just the right way produced a metal that was much harder than either one alone. This mixture of metals, or alloy, was bronze; its discovery delineates the beginning of the *Bronze Age* and the end of the Stone Age that had lasted for some 2.5 million years. Soon, craftsmen and artisans learned forging and casting techniques to make a wide array of objects, including sculptures, plows, wheels, sharp knives, axes, hammers, armor, and many types of weapons. The Bronze Age lasted for the next 1500 years, until the more difficult process of iron working and steel production was refined.*

The development of agriculture, bronze technology, and sedentary living in large communities was a new direction for humans that may be seen as progress, but it came at a price. According to historian John Coatsworth,

> Bioarcheologists have linked the agricultural transition to a significant decline in nutrition and to increases in disease, mortality, overwork, and violence in areas where skeletal remains make it possible to compare human welfare before and after the change. (Coatsworth 1996)

* Steel is an alloy of iron with a small amount of carbon or other metals added to improve hardness. The earliest known steel comes from a site in modern day Turkey and dates to about 4,000 years ago. Steel artifacts from 3400 year ago have been found in East Africa.

Given these drawbacks, we may question whether it was really a step forward for people to move toward agriculture and high-density city living, and away from the foraging lifestyle of hunting, fishing, and gathering. Yet the move may have been inevitable as human populations grew, because the foraging lifestyle simply could not supply enough food as natural resources became depleted. If population was to continue growing, a new way to produce food had to be invented. Agriculture could feed large numbers of people and this became a positive feedback loop: larger population required agriculture, and agriculture brought larger population, a trend that has continued to the present day. Even though agriculture could produce large and reliable *quantities* of food, the *quality* of a grain-centered diet was lower than the meat-based diet of hunter-gatherers.

Although the Neolithic revolution began in the Fertile Crescent, it later appeared independently in all three world zones. By 10,000 years ago, plants were being domesticated in northern Peru (Dillehay et al. 2007), by 9500 years ago horticulture was well established in the Huang He (Yellow River) valley of China, and by 9000 years ago it appeared in New Guinea, the Nile valley of Egypt, the Indus valley of Pakistan, and the Mediterranean coast of Europe. It's easy to understand how agriculture and technology spread throughout the Fertile Crescent and Eurasia where people moved about easily, but how did it develop almost simultaneously in such isolated locations as Peru and New Guinea? Also puzzling are those peoples who never embraced agriculture, but remained as hunter-gatherers. The Australian Aborigines, the Khoisan Bushmen of South Africa, and many of the indigenous tribes of North America all rejected agriculture, even though they had contact with farming peoples from nearby regions.

The Neolithic period brought the widespread transition from the nomadic hunter-gatherer lifestyle to a sedentary agricultural lifestyle from about 11,000 to 6000 years ago. During this time the world was a relatively peaceful * collection of village communities where people practiced both foraging and farming. Resources were still plentiful and the rate of population growth was still low by today's standards. There was extensive movement of people and exchange of goods, especially throughout the Fertile Crescent and Central Asia. The stage was now set for the final great development in our story, and this one would carry humans for the next 5000 years into the modern world. Mesopotamia, with its ideal climate, great rivers, fertile soils, and dense network of trade routes connecting proto-cities, was the favored location for the birth of civilization.

* Evidenced by the absence of weapons and the rarity of violent death based on skeletal remains.

Chapter 12
Consolidating Power

Civilization and Empire

The Neolithic revolution that began about 11,000 years ago brought with it agriculture, permanent settlements, and extensive trade to humans, but until about 7,000 years ago there was little social stratification or class rank in the communities of people. The best evidence of this comes from proto-cities like Catalhoyuk and Jericho, where there are no signs of significant differences in wealth: most houses were about the same size, there was little difference in the way the dead were buried, there were no monumental public buildings like temples or palaces, and the nutritional levels were similar across the whole population (skeletal remains show this). The social structure of these communities was probably *kin-ordered,* based on family lineages. People within a community saw themselves as related through common ancestors, and authority was based largely on age and seniority. While some family lineages may have had a higher status than others, power and authority were granted largely by consent; but all this was about to change.

The first evidence of a new social order is probably the temple of Eridu in Mesopotamia (present day southern Iraq) that dates to about 7000 years ago. The appearance of monumental architecture meant that there were powerful leaders or managers who coordinated the labor of thousands of workers—most likely slaves. Secular and religious power probably developed together, and wealth became concentrated in small elites. Monumental architecture, like the pyramids and ziggurats (magnificent stepped structures), was intended to inspire awe and fear of the gods and symbolize the power of the priests and rulers who claimed to represent the gods.

As cities grew and population densities increased, many social problems arose that could not be handled by simple consent. New social mechanisms had to be invented to settle disputes, organize trade, distribute food and water, dispose of wastes, and provide defense from attack. The need for central authority was both obvious and inevitable. However, central authority meant that power and resources became concentrated in the hands of a small ruling class. This brought a shift from power based on consent to power based on coercion. This consolidation of power was the beginning of a new social structure known as the city-state.

The emergence of the city-state is widely regarded as the beginning of civilization. The first city-states appeared by about 5200 years ago in southern Mesopotamia, a region also known as Sumeria. Eventually Sumeria grew to include twelve independent city-states, including Eridu, Uruk, Nippur, Ur, Kish, and Lagash. Archeologists and historians have traditionally viewed Mesopotamia, and Sumeria specifically, as the "cradle of civilization" but recent discoveries in Iran and Central Asia suggest that city-states emerged about 5000 years ago in many locations, and that multiple societies contributed to early civilization (Ur et al. 2007). While it was once thought that southern Mesopotamia, lacking mineral resources, was a single hub of trade, it now appears that there were many large trade centers throughout Central Asia. The development of agrarian (agriculture-based) civilizations and city-states seemed inevitable in other parts of the world as well: over the next thousand years they appeared independently in Egypt, Crete, China, India, Peru, and somewhat later in Mexico.

The emergence of city-states was closely related to improvements in agriculture. New technologies and techniques, like the plow, the wheel, large-scale irrigation, and the use of animals (including human slaves) increased production and allowed for larger and larger populations to be fed (on a low-quality diet, however, as noted earlier). However, the ruling elite took control of food production and distribution. Evidence of social hierarchy and inequality is abundant in all of the first city-states: houses vary greatly in size, elaborate burials were given only to the ruling class, and nutritional differences are clear in the skeletal remains. From now on—for the next 5000 years—life would be harsh and short for the common person, as the royalty lavished in luxury. Describing the emergence of state power and inequality, anthropologist Marvin Harris writes,

> For the first time there appeared on earth kings, dictators, high priests, emperors, prime ministers, presidents, governors, mayors, generals, admirals, police chiefs, judges, lawyers, and jailers, along with dungeons, jails, penitentiaries, and concentration camps. Under the tutelage of the state, human beings learned for the first time how to bow, grovel, kneel and kowtow. In many ways the rise of the state was the descent of the world from freedom to slavery. (Harris 1974)

Despite this dark view of state power, Harris was correct in pointing out that slavery emerged hand-in-hand with centralized power. Slaves were people at the very bottom of the social hierarchy who were owned like animals. Humans became an important source of energy that civilization depended on, in much the same way we depend on fossil fuels today. Slavery would be the engine of civilization for the next 5000 years, converting food into mechanical energy, making possible everything from the great pyramids of Egypt and Mexico to the vast plantation economy of the southeastern United States. The use of slavery would not be seriously questioned until the 1800s, at about the time other energy sources like coal and oil became available to power mechanical engines.

Along with the social hierarchies that developed within the city-state there arose male-dominated power structures, or *patriarchy*. The reasons for this are still debated, but the best explanation may be that males were less vital in the household, which was still at the heart of all societies. As division of labor evolved, it was the men who took on specialized, full-time occupations because the women were needed at home. Inevitably two worlds emerged: the household world occupied by women, and the public domain dominated by men. At the highest levels of power were usually men, and this would not change appreciably for the next 5000 years.

The development of states and concentrated political power gave rise to a massive accumulation of resources and wealth among the elite. It was the need to keep track of crops, livestock, slaves, and other possessions that brought about the first systems of writing. The earliest known writings, from Uruk about 5200 years ago, were simply inventory lists and records of how goods were received and distributed—*paleo bookkeeping*. Early written symbols were made by impressing a soft clay tablet with a reed stylus to create wedge-shaped symbols known as *cuneiform* writing. When the clay hardened, a permanent record existed, providing a way to store and distribute information. Recent discoveries at Jiroft (in modern Iran) and other Central Asian sites suggest that several writing systems developed in multiple locations throughout this part of the world at about the same time (Lawler 2007).

Figure 12-1
Chronology of Early Civilization

Event	**Approximate Date (years before present)**
End of last ice age, beginning of the Neolithic Age	11,500
First horticulture and domestication of plants and animals	11,000
First agriculture and proto-cities, like Jericho and Catalhoyuk	10,000
First true cities, like Eridu	7,000
First use of copper in Anatolia	7,000
Bronze Age begins in Sumeria, end of the Neolithic Age	5,500
First city-states in Sumeria, such as Uruk, Nippur, and Kish	5,200
Early city-states in Egypt and Iran	5,000
Early city-states in Northern India, Peru, and Crete	4,500
First empire (Babylon)	4,200
Early city-states in Northern China	4,000
Early city-states in Mexico	3,200

Over the next few thousand years writing became more sophisticated and appeared in all early civilizations including Egypt, the Indus Valley, China, Crete, and finally Mesoamerica. Along with writing, calendars based on the motions of the Sun and Moon were developed so that important events like the planting of crops could be regulated. Mathematics probably originated at about this time to help with the design, layout, and management of farmlands.

The elites who developed the first writing systems soon began using writing for more than just accounting—it became a way of permanently recording the knowledge of the civilization and controlling who had access to it. Before writing was developed, knowledge was passed along orally, through storytelling, and the information inevitably changed after many retellings. The written record brought a permanence and integrity to information and knowledge for the first time. This knowledge, however, was not accessible to everyone. The scribes themselves were a select few who were trained in the secrets of writing, while the ruling class hoarded the written records as a way to solidify their own power. For the next 5000 years, writing, and the knowledge it contained, was reserved for a small elite among people, and only in the last few hundred years has the written word become widely available to all people.

The emergence of writing about 5000 years ago also marks the beginning of recorded history. The study of written history as a formal academic field begins with the first writings of Sumerians, and continues through the civilizations of Babylon, Egypt, China, India, Greece, and Rome, who all left extensive writings. The first written history is attributed to Herodotus in Greece who authored accounts and commentaries on the Persian wars about 2500 years ago (490 BCE). The writings that have come down to us today reveal the essence of the society and culture, providing a glimpse into the ancient psyche; the Greek tragedies of Euripides dating to about 2500 years ago (420 BCE) show that we struggle with many of the same dilemmas and are afflicted with many of the same frailties now as then.

The transformation of the human condition from peaceful hunter-gatherers 11,000 years ago to sedentary city-dwellers universally brought another trend: conflict and war, resulting from the ownership of property and territory. Richard Leakey writes,

> The hunter-gatherer is part of the natural order: a farmer necessarily distorts that order. But more important, sedentary farming communities have the opportunity to accumulate possessions, and having done so they must protect them. This is the key to human conflict, and it is greatly exaggerated in the highly materialistic world we now live in. (Leakey 1991)

The first true cities that appeared in the Fertile Crescent 5000 years ago show clear signs that war was commonplace. Fortified walls around cities were universal, military equipment is abundant, and burials with weapons are common, while none of these things can be found in the proto-cities from just a few thousand years earlier. Armies were raised for the first time and served multiple purposes: they could maintain the power of the ruling elite by enforcing its wishes within the state, they could defend against invading peoples, and they could invade neighboring states in the quest for greater territory and wealth.

The emergence of the city-state and its centralized power signaled the beginning of tribute-taking societies. In exchange for protection, all common people had to pay some form of tribute, or taxes, to the rulers. At first, tributes probably consisted of food, labor, daughters, wives, animals, or land, but later came to include precious metals, artwork, and money. In earlier pre-state societies, resources were shared and exchanged largely by consent, but state power meant that resources could be extracted by force from the common people and amassed by a small elite. This trend would continue unabated for the next 5000 years until very recently when humans began to envision a new social order based on equality of opportunity, dispersed power, and freedom for everyone.

As tribute-taking societies grew in population, they had to acquire more resources to support their increasing numbers, and this meant expanding their sphere of tribute-taking by invading neighboring states. After centuries of war among the Sumerian city-states, they were eventually unified about 4200 years ago into the first true empire, centered in the city of Babylon. The Babylonian Empire would last for 1700 years, until it finally collapsed and disappeared, but the age of empire had begun, and many more would rise and fall, right up until the present day. The most well known of these include the Egyptian Empire (2500 BCE-500 BCE), the Persian Empire (550-330 BCE), the Roman Empire (44 BCE-395 CE), the Tang Empire (618-907), the Mongol Empire (1206-1368), the Ming Empire (1368-1644), the Ottoman Empire (1453-1566), the British Empire (1583-1914), the Axis Empire of Nazi Germany and Imperial Japan (1939-1945), the Soviet Empire (1946-1989), and the American Empire (1945, after World War II).

The Rest is History

We have finally arrived at the end of our story, but why should the story of how we got here end 5000 years ago? Certainly a great deal has happened since the time of ancient Sumeria and the Babylonian Empire. The first reason is because the time span of our creation story is almost 14 billion years, and on this scale the advent of civilization 5000 years ago is, for all practical purposes, the same moment as today. All of our glorious 5000 years of civilization fills only the last *one-tenth of a second* in the three geocosmic days that represent the life of the universe!

Another reason for ending our creation story 5000 years ago is that the events and players since then have been thoroughly documented and analyzed by historical scholars. The traditional field of history begins with the first civilizations and continues up to the present day. There would be very little new to add. However our *deep history*, the story of what happened *before* traditional history, is quite new; some scholars now call this *Big History.*[*] In this book we have explored our big history, and it's a new story that was not possible to tell in such detail just a few decades ago.

* A number of excellent works of big history have appeared in the last decade. See, for example, Christian, David. 2004. *Maps of Time: An Introduction to Big History*. Berkeley CA: University of California Press.

From the new perspective of big history, of seeing our story spanning 13.75 billion years, the last 5000 years takes on a different look. The people of Sumeria and Babylonia and Egypt and Greece and Rome were not that different from us today, at least compared to the Neanderthals and Cro-Magnons 35,000 years ago, who were using stone tools. The advent of civilization some 5000 years ago, and the age of empire that soon followed, marks the beginning of the present era of humanity, a world of cities and states and nations, of powerful rulers and taxation and wars, of dazzling technology, magnificent art, and monumental architecture. No doubt our technology is far more powerful and sophisticated today, and there are many more of us on the planet, but how much has really changed? Most of the things that characterized the beginning of civilization 5000 years ago are still with us today, while none of these things existed 8,000 years ago; these include:

- Large-scale agriculture, leading to reliable food production and large, permanent population centers.
- Highly structured cities that include streets, districts, multi-storied buildings, water delivery systems, public spaces, monumental architecture, and fortified enclosing walls.
- The division of labor and the rise of trades and professions.
- Political systems that consolidate power and concentrate resources into the hands of a small ruling elite who govern city-states and enforce tribute taking (taxation).
- Extensive social stratification and inequality of resources, with most wealth held by a small ruling elite.
- Male domination of public life and political power.
- A heavy reliance on human slaves as a source of mechanical power and productivity.
- Armies, military systems, and war.
- Expansion of territory, driven mostly by population growth, often leading to empires.
- Writing systems, used as a way to record, accumulate, and control knowledge.
- Systems of religion that are often part of the state power structure and often produce monumental architecture and other artistic works.
- Trade networks that allow states to acquire scarce resources, and that also become conduits for the spread of knowledge.
- Explosive growth of knowledge and technology: mathematics, astronomy, medicine, literature, metallurgy, and eventually empirical science.

It has been said that history repeats itself, and indeed the history of human civilization over the last 5000 years seems to be an endless string of wars and empires and all-powerful rulers and violence. Some would argue that we are doomed to repeat these things forever, because this is "human nature." However, this narrow perspective is based only on knowledge of the last 5000 years of recorded history, and until very recently that was all we knew about our past. Now, however, we have uncovered the much bigger story of how we got here from the Cosmos, and we can see the last 5000 years as a brief stage, a mere blink of the eye. We are now beginning to see that the story of humanity is far more than we thought just a hundred years ago, and that our story is still unfolding; there is no reason to think that we are stuck in an endless cycle of repeating history, or that we have arrived at our final destination. Rather, we should begin to see humanity as a work in progress, and to realize that there could be much more of our story that is yet to unfold.

Part Five Summary

By about 200,000 years ago the hominid lineage in Africa had given rise to anatomically modern humans—*Homo sapiens*. At this time, at least two other hominid species still roamed the Earth: *Homo erectus* in Asia and *Homo neanderthalensis* in Central Asia and Europe. As *Homo sapiens* flourished, however, both of these rivals would eventually go extinct, like every hominid species before them. After a seven million-year run, every experiment in the hominid line eventually led to extinction, except one: humanity. We are the last hominid standing.

The first humans in Africa looked physically very much like us today, and they had a full-sized brain. Their behavior and intelligence, reflected in the well-preserved tools they left behind, were still comparable to the Neanderthals. By about 100,000 years ago humans had spread over much of the African continent in small isolated groups, and a few began to leave Africa. Some got as far north as modern day Israel, and others traveled east into the Arabian Peninsula, and possibly as far as India and China.

Apparently none of these people who first left Africa survived in the long term, or at least their genetic material is nowhere to be found today. They were probably victims of the devastating eruption of Mt. Toba 73,000 years ago and the aftermath of volcanic winter that lingered for thousands of years; and those who had remained in Africa were almost eliminated as well. After struggling for survival against a massive drought for tens of thousands of years, they were further pushed toward extinction by the global volcanic winter from Toba. Analysis of DNA from living humans suggests that by about 70,000 years ago the total population of humans on Earth plummeted to perhaps only a few thousand who lived in Africa.

These few human survivors probably lived in two completely separate areas: some in what is now Ethiopia, and others near the coast of South Africa. These people became the founding population of all of humanity today, with almost all of us descending from the Ethiopian group. Just after 70,000 years ago, as conditions in Africa improved, the human population began to grow, and by about 60,000 years ago humans once again ventured out of Africa. This time, however, nothing could stop them and they spread over the entire Earth relatively quickly. The survivors of the Toba bottleneck were somehow different—a phenomenal change had occurred in the human brain that launched our species into a completely new realm of existence, unlike anything that had happened before on Earth. This change has been variously called the great leap forward, the big bang of brain development, and the dawn of human culture. No one knows exactly what happened in the brain or what caused it, but there is strong evidence suggesting that people 60,000 years ago were substantially different from people 100,000 years ago.

In the time since this transformative leap was completed, in the last 60,000 years, humans have evidently not evolved *biologically*. That is, our physical form, including the brain, is about the same. Yet a new and much more rapid kind of evolution has been taking place: the evolution of culture and consciousness. Humans evolved sophisticated culture and a social intelligence based on spoken language that gave them the ability to create and share new knowledge, and build a knowledge base that could be passed along from one generation to the next. With this new domain of intelligence, humans quickly spread over all of the habitable continents, completely replacing *Homo neanderthalensis* and *Homo erectus*, and driving many large mammals to extinction as well. *Homo sapiens*, the last hominid, wielded unprecedented power.

By the time the last ice age was ending about 11,500 years ago, humans lived all over the Earth in relatively peaceful, nomadic tribes as hunter-gatherers. Certain areas, like the Fertile Crescent in Central Asia, attracted people because of the mild climate and abundant plant and animal resources. A new way to acquire food began to emerge, based on protecting and nurturing certain plant and animal species. The domestication of plants and animals brought the end of the nomadic hunter-gather lifestyle that had served our ancestors for millions of years. The large-scale cultivation of crops boosted food production so that permanent populations could be supported; people began to live in large, disorganized proto-cities. As population density increased, the social dynamics became more complicated; inevitably governance and social hierarchy began to evolve.

By about 5200 years ago, city-states began to appear, at first in Central Asia and later in Egypt, India, Crete, China, and Mesoamerica. This was the beginning of civilization. Despite the widely separated locations of early civilizations, they were remarkably similar. And many of the social structures of early civilization that were in place by 4000 years ago are still with us today: we are still living in the era of civilization and empire that began 5000 years ago. Now, however, there are signs that some of these 5000-year-old trends are crumbling: slavery is widely regarded as unacceptable, patriarchy is on the decline, the effectiveness of military action is now in question, and the experiment of liberal democracy promises to give power and freedom to all people instead of just a few.

Epilogue

This is our creation story. It is a story that was largely unknown just fifty years ago; it's a story whose details are still being fleshed out as scientists make new discoveries. Now we should step back and reflect on it. What does this story mean for us today? What do we make of the long, improbable succession of events that culminated with modern human beings: from quarks and electrons that assembled into ordinary matter, to the magnificent stars and galaxies that emerged all over the universe, to a perfect planet with liquid water and a rich atmosphere where DNA could form and begin its relentless replication, to the evolution of life that continued for almost 4 billion years to finally produce modern humans, to the near extinction of our species, and finally the overpopulation of Earth? Did this happen by mere coincidence, the chance sloshing together of particles and chemicals that was bound to happen somewhere in our universe? Or, does the universe naturally evolve toward complexity and intelligent life, driven by some principle or force that science cannot yet identify? Does life exist elsewhere in the universe and, if so, have civilization and culture emerged anywhere else? What about our place in the universe? Are we important, or merely lucky? Are we insignificant specks of matter in a vast cold universe, or the culmination of nearly 14 billion years of evolution? And what does our new-found creation story tell us about the future of humanity and the civilization project? Have we reached our zenith and now must accept decline, and perhaps extinction, or are we transitioning into something new?

At present, these questions are beyond the reach of science, and therefore beyond the scope of this book, which is based on widely accepted science. Nonetheless they are some of the most important questions we can ask at this point in our story, and I hope to explore them in another book.

Appendices

Appendix I
Finding the Age of the Universe

Early Steps

Throughout this book, and particularly in Chapter One, we state that the universe is 13.75 billion years old; that is, the universe was born 13.75 billion years ago. How do we know this? This Appendix recounts the 400-year story of how we discovered the age of the universe.

Galileo took the first big steps in 1609 when he built his own small telescope (it had just been invented) and pointed it upward at the night sky. This was the birth of modern astronomy and a moment when our understanding of the universe and our place in it began to change drastically. Over the next two years, Galileo mapped the surface of the Moon in detail and discovered the moons of Jupiter, the phases of Venus, the Sunspots, and a few fuzzy patches in the night sky that came to be called *spiral nebulae*. The spiral nebulae remained one of the great, unexplained mysteries of astronomy for over 300 years, until Edwin Hubble solved that mystery in the 1920s. As it turned out, the spiral nebulae held the key to finding the age of the universe.

From the time of Galileo until today, progress in astronomy has been driven by improvements in the telescope (see *Exploring Deeper*, next page). As telescopes got bigger and better, astronomers could look further and further out at the exotic zoo of objects that fills the cosmos. Starting in the 1700s they began to catalog hundreds of stars with names and locations, and also to discover other types of objects: there were double and triple stars, giant star clusters, comets, including some with multiple tails, the moons of Saturn, the new planet Uranus, the polar ice caps of Mars, and many more of the fuzzy, mysterious spiral nebulae.

These early astronomers in the 1700s and 1800s were simply observing and cataloging the objects in the night sky, but they had no idea what they were seeing or how far away these things were. They could not even explain what a star was, something that would not become possible until nuclear physics was developed in the 1930s. They did not have any kind of explanatory model for the universe, and had no sense of its size.

This began to change in 1838 when Wilhelm Bessel made the first successful distance measurement to a star. Using the parallax technique (explained in Appendix II), he measured the distance to a nearby star called *61Cygni*, and found it to be about 60 trillion miles! In modern terms this distance is about 10 light-years, which means that light from this star takes ten years to reach us. Bessel's measurement was a startling discovery because it hinted for the first time at how VERY BIG the universe was, and what a powerful source of energy stars must be if we could see them at such great distances. This also deepened the mystery of what stars were, because there was no known source of energy (such as fire) that could produce even a tiny fraction of the energy stars produced.

In the time since Bessel first measured the distance to a star, many other star distances have been determined, and a number of other techniques for measuring star distances have been developed (see Appendix II). This allowed astronomers to construct a map of the local universe, much as the early European explorers like Magellan constructed a map of the surface of Earth. Until the 1920s, two questions lingered: how big is the universe and what were the spiral nebulae?

Exploring Deeper

The Evolution of Telescopes

For the last 400 years, astronomers have tried to make bigger and better telescopes so they could gather fainter light from more distant sources. Collecting light from further out in the universe is the dream of all astronomers because they are seeing the large-scale structure of the universe, which helps them make a better map of our place in the Cosmos; and they are also looking back in time.

All of the early telescopes were *refracting* telescopes – they used glass lenses to bend and concentrate light. However, the refracting telescope was limited in how much it could be improved—it could only be made longer, because larger diameter lenses were difficult to make and also contained imperfections that distorted the light. By about 1650, Hevelius built a refracting telescope that was nearly 150 feet long (Figure A-1). It was so unwieldy that it could barely be moved to look at different objects and would bend under its own weight.

Figure A-1

The best telescope on Earth in 1650

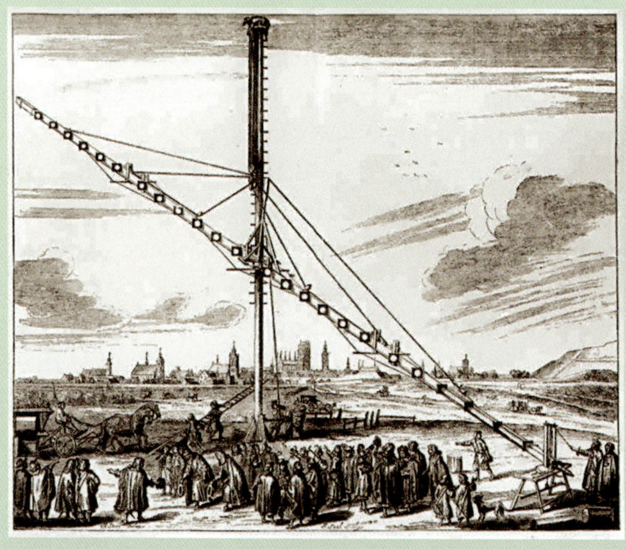

Wikimedia Commons

The best telescopes today are *reflecting* telescopes; they use a curved mirror instead of glass lenses to collect faint light and concentrate it. Isaac Newton built the first reflecting telescope in 1668, but they did not come into widespread use for almost 100 years because it was so difficult to construct high-quality curved mirrors. By the 1800s, astronomers started making bigger and better reflecting telescopes, and this continues today. In 1948 the Hale Telescope on Mount Palomar in California was built with a mirror diameter of 5 meters (almost 17 feet across), and today the largest optical telescopes are the Keck Telescope in Hawaii with a 10-meter collecting mirror, and the Gran Telescopio in the Canary Islands with a 10.4-meter mirror (34 feet across!).

Not all modern telescopes are optical telescopes that gather visible light. Some are built to detect non-visible wavelengths like radio waves, microwaves, ultraviolet radiation, or gamma rays. The largest radio telescope today is the Arecibo telescope, a huge dish suspended over a natural depression in a mountain in Puerto Rico. With a collecting diameter of 1000 feet, it can gather faint radio emissions from distant objects, painting a very different picture of the universe from what we see with visible light.

The best telescope today is probably the Hubble Space Telescope. Even though its mirror is only 2.4 meters across, it is located 366 miles above the surface of the Earth in a low orbit (Figure A-2). The advantages of being above Earth's murky atmosphere far outweigh the smaller mirror size, and since its deployment in 1990 the Hubble Telescope has revealed countless new facets of the universe. Its life is expected to come to an end around 2020 when the orbit decays and it falls to Earth, but a new and even better space telescope—the James Webb Space Telescope—is being designed and will be placed in orbit in 2013.

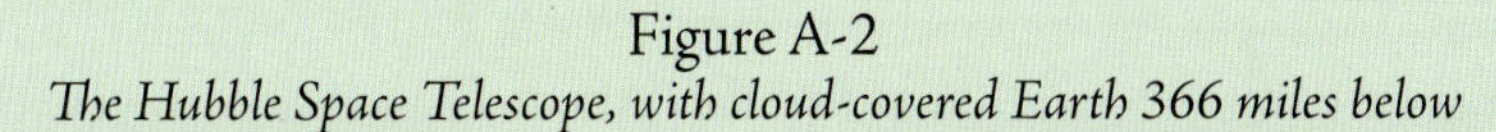

Figure A-2

The Hubble Space Telescope, with cloud-covered Earth 366 miles below

Courtesy: NASA/STSci

How Big is the Universe?

By about 1915 astronomers had measured distances to thousands of stars and found that they all seemed to lie within about 50,000 light-years. They also found that all of the known stars resided in a huge, flat, revolving disk that came to be called the Milky Way Galaxy. The name "Milky Way" comes from the hazy, luminous band stretching across the night sky that had been known since antiquity. This band outlines the plane of the galaxy, hinting at its flat shape.

But was this the entire universe? That is, did the universe consist of just the Milky Way Galaxy, a giant revolving collection of stars, or was the Milky Way just a small part of a much bigger universe? The answer to this very big question was to be found in the still-unexplained fuzzy patches that Galileo had seen 300 years earlier—the *spiral nebulae*. What were they and how far away were they? None of the methods for measuring the distance to stars could be applied to the spirals because they were not stars. The mystery of the spiral nebulae and the question of the size of the universe became the central focus of astronomy in 1920, and two schools of thought had emerged.

One view, championed by Harlow Shapley, was that the spiral nebulae were relatively small collections of stars that resided within the Milky Way. He had perfected distance-measuring techniques that showed that the Milky Way had a diameter of about 300,000 light-years and he believed that this was the *entire* universe.

The other view was first put forward by the philosopher Emanuel Kant in 1755, and taken up by Heber Curtis by 1920. Kant had proposed that the spiral nebulae were "island universes"—vast collections of stars like our own Milky Way, at very great distances far beyond our own galaxy. Curtis believed that the Sun was at the center of a relatively small Milky Way—only about 30,000 light-years across—but it was one of many island universes, all in a vast universe. In Curtis' view the spiral nebulae, or island universes, were actually other galaxies, much like our own Milky Way.

The question, then, was about the size and nature of the universe. Did the universe consist of a single revolving galaxy, the Milky Way, that was a few hundred thousand light-years in size, as Shapley believed, or was it much larger, perhaps many millions of light-years across and made up of many galaxies, as Curtis believed?

The debate over these two views came to a climax on the evening of April 26, 1920 when Shapley and Curtis each spoke at the Smithsonian Museum in Washington, D.C. during the annual meeting of the National Academy of Sciences. This event has come to be known as the "Shapley-Curtis Debate", but the idea of it being a formal debate, or a struggle between the two men, was probably a drama created by the press. More accurately, it was science at its best—two scientists with opposing views respectfully presenting their evidence and making their best case before their peers. It was an unprecedented moment, when one of the biggest questions of all was being considered: How big is the universe?

By most accounts, Shapley was more organized and more persuasive than Curtis, but no real conclusion was reached at the end of the evening. The final resolution rested on one question: How far away were the spiral nebulae? If the distances to spiral nebulae were less than a few hundred thousand light-years then they must reside *within* our galaxy, supporting Shapley's view. If the distances were much greater—perhaps in the millions of light-years—then they must be well *outside* our galaxy, as Curtis believed. However, in 1920 no one could measure the distance to a spiral nebula, so the issue could not be resolved.

The answer finally came three years later with the work of Edwin Hubble. Hubble was photographing Andromeda, the brightest of the spiral nebulae, with the new 100-inch telescope at Mount Wilson Observatory, hoping to measure its distance from Earth. With better light-gathering power than anyone before, he was able to detect variable stars in Andromeda, and using Henrietta Leavitt's technique (Appendix II) he successfully measured the distance to Andromeda. His initial result showed that Andromeda was about *one million* light-years away, putting it well outside the Milky Way Galaxy. If this was true, it meant that Andromeda was in fact a separate galaxy, as Curtis believed. Soon other measurements confirmed this, though Hubble's initial distance measurements were found to be short by a factor of about three because he could not account for the pervasive dust between stars in our galaxy that diminished the light intensity. Adjusting for this, we know today that Andromeda is about 2.6 million light-years away from us and is our sister galaxy. The Milky Way and Andromeda galaxies are gravitationally bound, circling each other, surrounded by many smaller satellite galaxies. We now know that we live in a cluster of galaxies which astronomers call the *Local Group*.

In the next few years, Hubble and others began the study of galaxies, opening a new era of astronomy and changing the map of the universe drastically. By the late 1920s we knew that our Sun (and Solar System) revolves around the center of our galaxy, the Milky Way, and that our galaxy was one of many galaxies in a vast universe that was at least *hundreds of millions* of light-years in size. But things were just getting interesting.

The Expanding Universe

There was another curious fact about galaxies that had been observed by Vesto Slipher and others since about 1914: the light from galaxies was almost always "stretched out", or *redshifted*. This could be seen when the faint light from a distant galaxy was collected in a large telescope, then dissected into its component wavelengths. This kind of spectral analysis had been used to determine the composition and surface temperature of stars, but now it revealed the redshift of galaxies.

Astronomers were familiar with redshifts and blueshifts in the light from stars within our own galaxy. These shifts in wavelength are caused by the *Doppler Effect*, which is familiar to us in sound waves. When a source of light, sound, or any kind of waves, is moving away from an observer the waves get stretched out. In sound waves this can be heard as a drop in pitch when a car drives by with its horn blaring. Similarly, when the source of waves is moving toward the observer, the waves are compressed. Astronomers have come to call the stretching of waves "redshifting" (red connoting longer) and the compressing of waves "blueshifting" (blue meaning shorter). The amount of stretching or compressing of the wavelength can be used to calculate the velocity of the moving source (Appendix II).

The technique of using the Doppler Effect to measure velocities is so effective that police rely on it today using radar—a type of radio wave—to measure the velocity of speeding cars. As a tool in astronomy, it is used to measure the motions of stars within our galaxy, and it helped establish that the galaxy rotates. When we look at stars *within* our galaxy we see both redshifts and blueshifts, indicating that some stars move away from us, and some toward us as they all revolve around the center of the galaxy.

However, Hubble saw something quite different in the light from distant galaxies. When he looked at galaxies outside our Local Group he saw *only redshifts*. This meant that these galaxies were all moving *away* from us. When Hubble calculated the speeds of these galaxies, it was astounding—thousands of miles per second! In every direction he looked, galaxies were flying away from us at enormous speeds. The universe, at the very large scale, seemed to be flying apart.

Through the late 1920s Hubble and his colleague Milton Humason measured the distances to many galaxies and compared each distance to the amount of redshift in the light from the galaxy. They found that the further away a galaxy was, the greater the redshift; and the greater redshift meant a greater velocity of recession. In other words, the *further* away a galaxy was, the *faster* it was moving away from us. When they made a graph of their data, a remarkably straight line resulted (Figure A-3). As telescopes continued to improve, astronomers could look further and further out, and the simple linear relationship between distance and velocity of recession (now called *Hubble's Law*) was still found to be true. Today we have observed galaxies that are 12 *billion* light-years away, receding at speeds close to the speed of light, and the Hubble Law still seems to hold at these great distances (Figure A-4).

When Hubble published this result in 1929, it was one of the most startling revelations *ever* about the world we live in: our universe is flying apart at staggering speeds, expanding dramatically, even though this is not apparent in our everyday lives on Earth or within our galaxy. In less than ten years during the 1920s, our view of the universe had changed from a solitary galaxy with a size of about 100,000 light-years, to Hubble's vast cosmos, at least ten thousand times bigger, filled with galaxies flying apart at enormous speeds.

Figure A-3
Edwin Hubble's graph from his 1929 paper,
"A Relation Between Distance and Radial Velocity Among Extra-Galactic Nebulae"

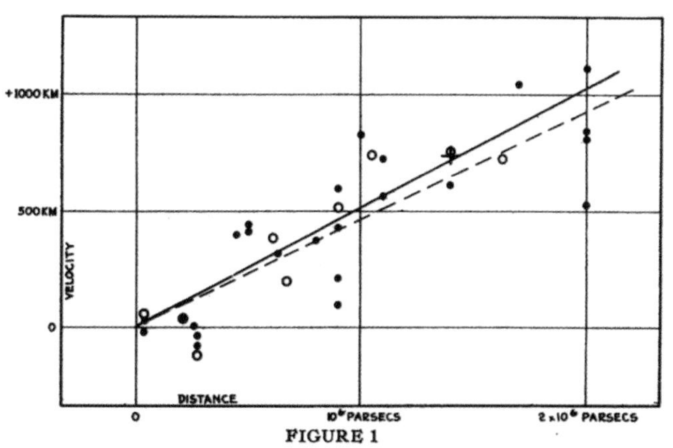

FIGURE 1

Velocity-Distance Relation among Extra-Galactic Nebulae.

Radial velocities, corrected for solar motion, are plotted against distances estimated from involved stars and mean luminosities of nebulae in a cluster. The black discs and full line represent the solution for solar motion using the nebulae individually; the circles and broken line represent the solution combining the nebulae into groups; the cross represents the mean velocity corresponding to the mean distance of 22 nebulae whose distances could not be estimated individually.

From *Proceedings of the National Academy of Sciences,* Volume 15 (1929)

Figure A-4
Typical modern data showing the recession velocity of galaxies versus distance away.
Each dot on the graph represents one galaxy.

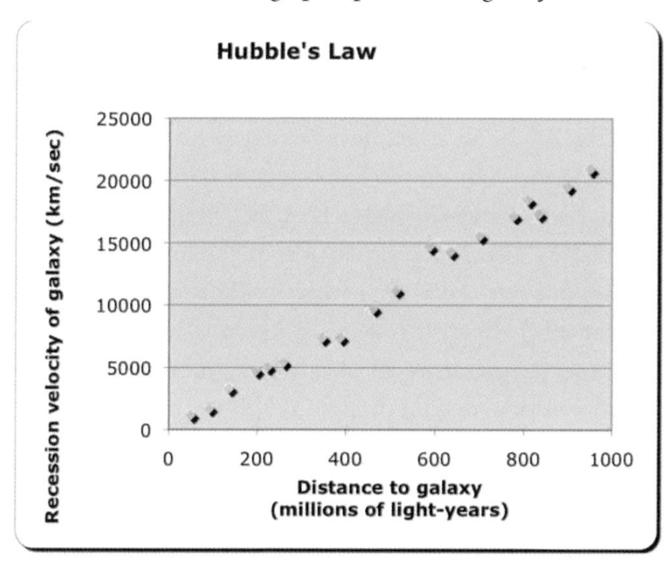

The Big Bang

What does this mean, that the galaxies are flying apart and the universe is expanding? By 1915, Einstein had found mathematically that his new General Theory of Relativity predicted a universe that could either expand or contract. He could not believe this was possible so he rejected the idea by inventing a factor in his equations that kept the universe static and unchanging. Others, like Alexander Friedmann and Georges Lemaitre working in the 1920s, used the equations of General Relativity to support the idea of an expanding universe and suggested that this expansion implied a birth of the universe in one violent moment long ago—a "'big bang." When Hubble and others discovered the redshifts of galaxies, the expansion of the universe was undeniable. This seemed preposterous, and it still does, since the world we see does not seem to be flying apart. Gravity masks this expansion within our own galaxy and group of galaxies, and only when we look out at great distances can we see it.

The idea that the universe was born in a gigantic explosion was not immediately accepted. In 1948 Fred Hoyle, Hermann Bondi, and Thomas Gold proposed an alternative explanation for the expansion of the universe. Their *Steady State Theory* took the view that the universe was uniform in space and time; that is, it looked the same from every location, and had no beginning or ending. A universe with no beginning had a certain philosophical appeal—it disposed of questions about how it was created and what was happening *before* the beginning. To remain in this static or steady state, new matter would constantly be created to compensate for the out-flowing expansion. Their calculations showed that new matter would need to appear at a modest rate of about one atom per year in a 100-meter cube of space. Hoyle and the other adherents could not account for how this new matter was created or where it came from—apparently it would just appear in empty space—but the effect was so subtle that it *could* be happening without notice.

Meanwhile, the *Big Bang Theory* was being refined. Theorists like George Gamow and Ralph Alpher, working in the late1940s, surmised that the very early universe, just after the Big Bang, must have been *very* hot in order to produce the first hydrogen and helium, and that an intense burst of light would have been produced shortly after the Big Bang. They suggested that this radiation should still linger today as microwaves, stretched out by the expansion of the universe. These early Big Bang theorists predicted that a pervasive microwave background radiation should fill all of space today, and that this *cosmic microwave background* would give the entire universe an overall background temperature of about 5 degrees above absolute zero. If this microwave remnant could be detected, it would be strong evidence that the Big Bang had actually occurred.

By the early 1960s the search was on for the cosmic microwave background radiation. Robert Dicke and James Peebles began building large sensitive detecting antennas at Princeton that could listen for this faint remnant of the Big Bang. While they were not having any success, just a few miles away at Bell Telephone Laboratories Arno Penzias and Robert Wilson were experimenting with a new type of radio telescope that was sensitive to microwaves. They could not get rid of a faint background hiss that remained no matter where they pointed their antenna. When they heard about the work of Dicke and Peebles, they realized that they must have detected the cosmic microwave background radiation. Their measurements, and others that soon followed, showed that the universe has a background temperature of 2.7 degrees, not far from Gamow's prediction of 5 degrees.

Penzias and Wilson were the first to detect the redshifted aftermath of the Big Bang—the cosmic microwave background—and won the Nobel Prize in 1978 for this work. After their discovery was published, the Big Bang Theory became widely accepted and the Steady State Theory largely disappeared. Since the 1960s, the evidence supporting Big Bang cosmology has accumulated. Since the 1990s, the microwave background has been studied with incredible precision, revealing details about the early universe just after the Big Bang.

Exploring Deeper

Are We at the Center of the Expanding Universe?
No matter what direction we look, we see galaxies flying away from us; and the further away they are, the faster they are receding from us. Doesn't this suggest that we are at the center of the universe? Cosmologists have developed several analogies to help people understand the expansion of the universe and why the expansion seems to be centered on Earth.

A two-dimensional analogy: the inflating balloon. Imagine a round balloon just slightly filled with air. While it is small let's draw dots all over the surface of the balloon—these will represent galaxies. Now, let's watch what happens to these dots as the balloon gets inflated with air and grows in size. Of course, the dots get further and further apart as the surface of the balloon expands. Now suppose we could place an observer on one of the dots and have her watch the other dots as the surface expands. No matter where she looks, every dot is moving away from her, and she could be on *any* dot and see the same thing. No one dot is at the center of the expansion because it's really the space between the dots—the surface of the balloon—that is expanding. With real galaxies in the universe, it is likewise the *space between them* that is expanding.

Another conclusion can be reached by our observer on one of the dots on the expanding surface of the balloon. If she looks at a neighbor dot that is very close, it moves away slowly, but if she look at dots further out, they move away faster. The more expanding space there is between dots the faster they move apart. This is exactly what Hubble discovered about the galaxies we observe from Earth—the further away they are, the faster they move away from us.

A three-dimensional analogy: the baking raisin bread. Now let's imagine some bread dough with raisins in it that we will bake into a loaf of bread. As it bakes, the bread dough "puffs up" and expands to a much larger size. In this analogy, the raisins represent galaxies and the expanding bread dough represents the expanding space between galaxies. Let's again put our observer on one of the raisins and imagine what she sees as she looks out at other raisins. No matter where she looks, she sees the other raisins moving away from her, so she might conclude that she is at the center of the expansion. However an observer on any raisin would see the same thing; and the further away a raisin is from an observer, the faster it would be moving away because there is more expanding bread in between.

Astronomers, starting with Hubble in about 1930, realized it should be possible to date the Big Bang by knowing the expansion rate of the universe. That is, we should be able to play the expansion backwards and find out when the expansion began—this would be the age of the universe. This expansion rate is of great interest to astronomers and is now called *Hubble's Constant*. It can be determined by finding the slope of the graph of velocity of recession versus distance (Figures A-3 or A-4). If we use Hubble's Constant to play the expansion backwards, we can find out how long ago all the galaxies were theoretically at the same point—*the beginning*. This amount of time is the age of the universe and is given by the inverse of Hubble's Constant (this is shown in Appendix II).

Yet determining the actual value of Hubble's Constant, or the expansion rate of the universe, is not an easy thing. Our ability to measure distances to galaxies has steadily improved since Hubble first did it in 1925, and likewise the value of Hubble's Constant has improved (see Figure A-5 below). When Hubble first estimated the age of the universe, he got about 2 billion years, but he knew this could not be right because there was already evidence at this time that *Earth* was older than this (see Appendix III). Over the next seventy years, estimates of the age of the universe ranged from 10 billion to 20 billion years. Finally in 2003, with detailed observations of the cosmic background radiation, scientists were confident that they had found an accurate value for Hubble's Constant and the age of the universe.

Today the most widely accepted value of Hubble's Constant is written as 71 kilometers per second per megaparsec, a strange sounding number that describes the rate of expansion of the universe (we could also express this as 22 km/sec per million light-years). Hubble's Constant tells how the velocity of recession of galaxies increases with distance, as a careful look at the units will show. A higher rate of expansion (a larger value for Hubble's Constant) implies a younger universe—it did not take as long to reach its present state—while a slower expansion rate means the universe is older. If the value of Hubble's Constant is 71, as most astronomers now accept, the Big Bang happened 13.75 billion years ago.

We have found the age of the universe!

Figure A-5
Variation in the value of Hubble's Constant since 1930

Value of Hubble's Constant (in km/sec per megaparsec)	Corresponding Age of the Universe (the inverse of Hubble's Constant)
500 (Hubble's first estimate)	2 billion years
50	20 billion years
100	10 billion years
71 +/- 3 (today's value)	13.75 +/- 0.11 billion years*

* For a complete listing of research findings from the WMAP mission, and conclusions about the age of the universe, see the NASA website: http://lambda.gsfc.nasa.gov/

Appendix II
The Tools and Techniques of Astronomy and Astrophysics

A. *Parallax: Measuring Distances to Nearby Stars*

The *parallax technique* makes use of the same principle that gives us depth perception with our eyes: when we view an object from two slightly different perspectives (our two separate eyes) we can tell how far away the object is. We have almost no depth perception with just one eye, and also very little depth perception if objects are too far away (we cannot tell much difference between something 10 miles away and 11 miles away).

In astronomy the two slightly different perspectives are two opposite sides of Earth's orbit around the Sun (Figure A-6). An astronomer can point her telescope at a star of interest and make an accurate determination of the exact direction to the star along the line of site. If she looks at the same star six months later, when Earth has moved to the opposite side of its orbit, the line of site to the star will shift slightly—this is called the parallax shift. If she can measure the very tiny angle of the shift in the line of sight, then the distance to the star can be determined. This can only be done for nearby stars (within about 1600 light-years) but so far we have cataloged several *hundred thousand* stars within this distance (with the help of the Hipparcos satellite)!

Figure A-6

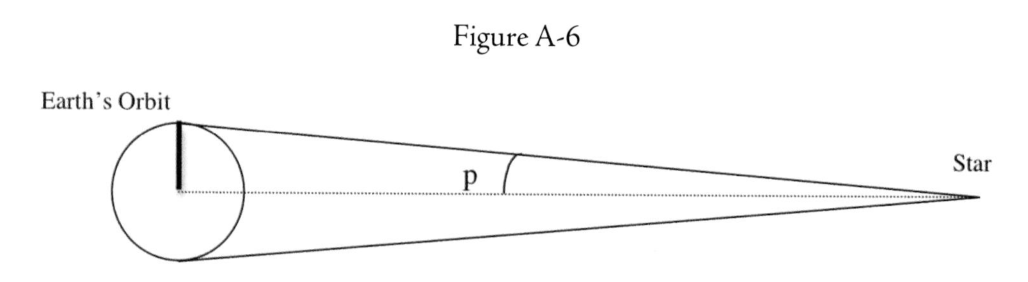

The *parallax equation* is a very convenient way to calculate the distance to the star:

D = 1/p where **D** is the *distance* to the star in *parsecs* and **p** is the *parallax angle* measured in *seconds of arc* (also called *arcseconds*). One arcsecond is 1/3600 of a degree. Notice in the diagram how the parallax angle is defined.

This equation also defines the *parsec* because D = 1 parsec when p = 1 arcsecond. This means that 1 parsec is the distance at which the radius of Earth's orbit subtends one arcsecond. One parsec is equal to 3.26 light-years, and this is also (conveniently) about the average distance between stars in our neighborhood of the galaxy (which turns out to be a fairly quiet, almost lonely, neighborhood).

Example: An astronomer measures a parallax shift of 0.02 arcseconds as she observes a star from opposing sides of Earth's orbit (6 months apart). What is the distance to the star in (a) parsecs? (b) light-years? (c) meters? (d) miles?

Solution: (a) From the diagram, notice that the parallax angle p is *half* the parallax shift of 0.02 arcseconds that was measured as the Earth moved from one side of its orbit to the other side, so p = 0.01 arcsec

Then D = 1/p = 1/(0.01arcseconds) = <u>100 parsecs</u>

(b) (100 parsecs) x (3.26 light-years/parsec) = <u>326 light-years</u>

(c) (325 light-years) x (9.48 x 10^{15} meters/light-year) =

<u>3.08 x 10^{18} meters</u>

(d) (3.08 x 10^{18} meters) x (1 mile/ 1609 meters) =

<u>1.91 x 10^{15} miles</u>

Note 1: Notice that the distances in meters and miles are both unwieldy numbers, while the distances in parsecs and light-years are much more manageable. The light-year is probably the most user-friendly and familiar unit for measuring distances to stars.

Note 2: The smallest parallax angle that has been measured by astronomers is about 0.002 arcseconds (the Hipparcos satellite did this), which corresponds to a distance of 500 parsecs. This means that the parallax method can be used for star distances *up to* about 500 parsecs or 1625 light-years, but no further. These are "nearby" stars.

B. The *Inverse-Square Law* for light can be written in two equivalent forms:

<u>Physics form:</u> $\mathbf{I = S/4\pi D^2}$

where **I** is *intensity* in watts/m^2, **S** is *source strength* in watts, and **D** is the *distance* to the light source in meters.

Astronomy form: $\mathbf{b = L/4\pi D^2}$

where **b** is *apparent brightness* and **L** is *luminosity*. Notice that if the apparent brightness and luminosity are known, this equation can be used to find the distance to a star **D**.

Example 1: A man observes a distant light bulb on a dark night. When he points his light meter at the bulb, the meter reads 8.0 x 10^{-6} watts/m^2. If the bulb has a source strength (or luminosity) of 100 watts, how far away is it?

Solution: Solve the inverse square law (let's use the astronomy form) for distance:

$D = (L/4\pi b)^{1/2} = (100/4 \times 3.14 \times 8.0 \times 10^{-6})^{1/2} =$ 998 meters **or** 0.62 miles away.

Example 2: An astronomer observes a distant star on a dark night. The star's apparent brightness, as measured by her telescope, is 2.5 x 10^{-14} watts/m^2. If the star's luminosity is known to be 7.5 x 10^{27} watts, (Section C explains how this can be known.) how far away is it?

Solution: Use the inverse square law as in example 1:

$D = (7.5 \times 10^{27} / 4 \times 3.14 \times 2.5 \times 10^{-14})^{1/2} =$ 1.55 x 10^{20} meters or 16,302 light-years away.

C. The *Stefan-Boltzmann Law* gives the luminosity of hot objects (like stars):

$\mathbf{L = 4\pi R^2 \sigma T^4}$ where L is the luminosity in watts, R is the radius of the object in meters, σ is a constant (like π) called the Stefan-Boltzmann constant with a value of 5.67 x 10^{-8}, and T is the surface temperature of the object in Kelvin degrees.

The most important thing this law says is that the luminosity of a star (how brightly it shines) is given by the *fourth power* of its temperature. This means that if one star has 3 *times* the temperature of another, it will shine 3^4 or *81 times* more brightly.

Example: An astronomer observing a star with his telescope estimates its radius to be 1.4 x 10^9 meters and determines its surface temperature to be 10,000 degrees. What is the luminosity of the star? (Section D explains how surface temperature can be found.)

Solution: Use the Stefan-Boltzmann Law:

$$L = (4)(3.14)(1.4 \times 10^9)^2(5.67 \times 10^{-8})(10{,}000)^4 = \underline{1.4 \times 10^{28} \text{ watts}}.$$

This is about 40 times as bright as the Sun.

D. *Wien's Law* is used to find the surface temperature of a hot object (like a star) by analyzing the colors (or wavelengths) of light it produces:

$$\mathbf{T = (.0029)/\lambda_{max}}$$

where **T** is the *surface temperature* in Kelvin degrees and λ_{max} is the *wavelength of peak intensity* in meters.

The wavelength of peak intensity is found by breaking up the light from the object into its wavelengths, a process called *spectral analysis*. Any hot object, like a piece of red-hot iron, a star, or even the human body, radiates light in a bell-shaped curve. Physicists call this a *blackbody curve*, which is a graph of *intensity* versus *wavelength* (Figure A-7).

The important thing is that there is an obvious peak in the black body curve at a certain wavelength—this is the wavelength of peak intensity, or λ_{max}. For hotter objects λ_{max} is at a shorter wavelength (toward blue), and for cooler objects λ_{max} is at a longer wavelength (toward red). Wien's Law tells us the temperature of the object from λ_{max}.

Example 1: When astronomers analyze the light from the sun, they find that the predominant wavelength (the peak in the black body curve) is about 500 nanometers. What is the surface temperature of the Sun?

Solution: First convert 500 nanometers into meters,

$$500 \text{ nm} \times (10^{-9} \text{ meters/nm}) = 5.0 \times 10^{-7} \text{ meters},$$

then use Wien's Law,

$$T = (.0029)/\lambda_{max} = (.0029)/(5.0 \times 10^{-7} \text{ meters}) = \underline{5800 \text{ degrees}}.$$

It's amazing to think that we can find the surface temperature of a star!

Figure A-7
Blackbody curves for objects at 5000 K, 4000 K and 3000K.

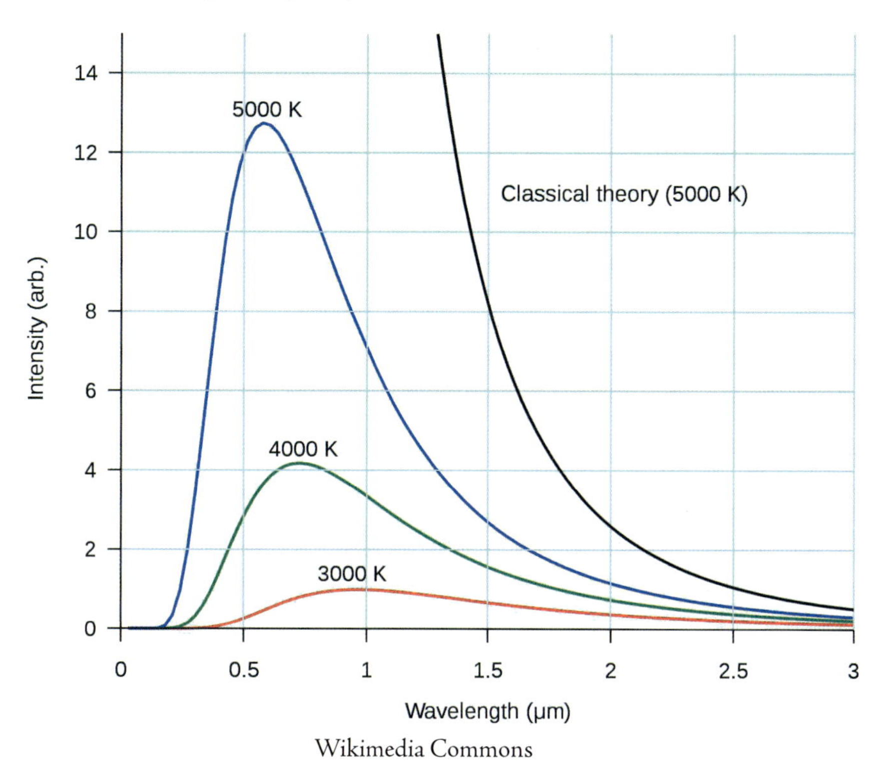

Wikimedia Commons

E. The *Spectral-Luminosity Method* uses Wien's Law, the Stefan-Boltzmann Law, and the Inverse–Square Law to find the distance to a star.

Example: Suppose an astronomer studies a star and determines from spectral analysis that its wavelength of maximum intensity is 400 nm, and then determines its radius to be 9.8×10^{8} meters, and measures its apparent brightness to be 4.2×10^{-13} watts/m^{2}. Find the distance to this star.

Solution: First, use Wien's Law with λ_{max} = 400 nm to find the surface temperature of the star. The answer is 7250 degrees.
Next, use this temperature in the Stefan-Boltzmann Law to find the luminosity. The answer is 1.9×10^{27} watts.
Finally, use this luminosity in the inverse-square law, along with the apparent brightness, to find the distance.
The final answer is 1.9×10^{19} meters or 2000 light-years.

F. *Measuring Distance Using Cepheid Variable Stars.* In the early 1900s, Henrietta Leavitt, working at Harvard College Observatory, developed an important new distance-measuring technique. She was studying variable stars, a type of star that regularly gets brighter and dimmer over a period of days. She noticed that the most luminous variable stars had the longest periods of variation. After observing hundreds of variable stars, she found a relationship between the luminosity of the star and its period of variation. Once the luminosity was known, the distance to the star could be calculated from the inverse square law. When she checked the distances given from this method against the distances given by the other two methods (A and E above), she got good agreement. Leavitt's period-luminosity relationship for variable stars was a critical piece of Hubble's pioneering work with galaxies in the 1920s, and is still widely used today

G. The *Doppler-Redshift equation* is used to find the recession velocity of galaxies:

$$\mathbf{z} = \Delta\lambda/\lambda = \mathbf{v}/\mathbf{c}$$

Where **z** is called the *amount of redshift*, $\Delta\lambda$ is the *change in wavelength*, λ is the *normal wavelength*, **v** is the *recession velocity* of the galaxy, and **c** is the *speed of light* ($= 3.0 \times 10^8$ meters/s).

When astronomers look at the light from distant galaxies, they see that it is stretched out or redshifted—every wavelength is longer by a certain percent, and this percent (or decimal) is called the amount of redshift. This redshift happens because distant galaxies are moving *away* from us at very high speeds (which we call the *recession velocity*). The Doppler-Redshift equation relates the amount of redshift in the light from a galaxy to the recession velocity of the galaxy.

Example: Hydrogen normally emits light with very specific colors (or wavelengths) that are well known from experiments on Earth—these are the *spectral lines* of hydrogen. One such line has a wavelength of 656 nanometers (a red-orange color that only hydrogen produces) and another is 486 nanometers (blue). Suppose an astronomer looks at the light from a distant galaxy and, after analyzing its colors, finds that all the expected wavelengths for hydrogen are present, but the wavelengths are all slightly longer: the normal 656 nm wavelength has been stretched to 663 nm and the normal 486 nm wavelength has been stretched to 491 nm. From this, what is the recession velocity of the galaxy?

Solution: First find the amount of redshift using either wavelength (only one is needed): Using the red line,

$$z = \Delta\lambda/\lambda = (663 - 656)/(656) = 7/656 = 0.01 \text{ or } \underline{1\%}$$

or using the blue wavelength, $z = (491\text{-}486)/(486) = 0.01$ or $\underline{1\%}$

These should be the same—that is, both wavelengths (and every wavelength in the light from this galaxy) is 1% longer. This is a 1% redshift.

The last part of the equation gives the recession velocity:

$$z = v/c \text{ or } v = zc = (0.01)(3 \times 10^8 \text{ m/sec}) = \underline{3 \times 10^6 \text{ m/sec}}$$

This galaxy, with a 1% redshift, is moving away from us at 3 million meters per second or almost 2,000 *miles* per second!

It is easiest to understand the Doppler-Redshift equation if you notice the v/c is a fraction: it's the fraction of the speed of light that the galaxy is moving away at, and it's equal to the amount of redshift. So a galaxy with a 1% redshift is receding from us at 1% the speed of light, and a galaxy with a 10% redshift is traveling at 10% the speed of light. Astronomers have seen galaxies with redshifts that are so high that they must be moving away at almost the speed of light! This also means that they are *very* far away, as we see in section H.

H. *Hubble's Law* relates the recession velocity of a galaxy to its distance away from us:

$\mathbf{v = H_o d}$ where v is the recession velocity, d is the distance away from us, and H_o is Hubble's Constant. Astronomers usually express the recession velocity in kilometers/second, and the distance away in megaparsecs. Using these units, the value of Hubble's Constant is about 71 km/sec/megaparsec (this is currently considered to be the most accurate value, and this number is of great interest as we see below). Note: A megaparsec is a million parsecs, or 3.26 million light-years, or 3.08×10^{22} meters.

Example 1: Using Hubble's Law and the Doppler-Redshift equation to find the distance to a galaxy: An astronomer discovers a galaxy whose light is redshifted by 50%. How far away is it?

Solution: The Doppler-Redshift equation tells us that this galaxy is receding at 50% of the speed of light or $(0.5)(3 \times 10^8$ m/sec$) =$ <u>1.5×10^8 m/sec</u>

Convert this to km/sec: $(1.5 \times 10^8$ m/sec$)(1$ km/1000 m$) =$ <u>1.5×10^5 km/sec</u>

Then put this velocity into Hubble's Law and solve for d:

$d = v/H_o = (1.5 \times 10^5$ km/sec$)/(71$ km/sec/megaparsec$) =$ <u>2,100 megaparsecs</u>

2,100 megaparsecs is about 6.8 billion light-years, which means that the light collected by the astronomer left the galaxy 6.8 billion years ago, before Earth and Sun even existed!

Example 2: Using Hubble's Constant to find the age of the Universe: How long ago did the Big Bang take place?

Solution: To see what we need to do, imagine two galaxies that today are a distance d apart and traveling away from each other at velocity v. Imagine running this backward so that the galaxies are approaching each other. We can calculate the time it takes for them to collide by $t = d/v$ [this is like finding the time for a car traveling 60 miles/hour to travel 180 miles: $t = (180$ miles$)/(60$ miles/hour$) = 3$ hours]. In the case of our galaxies, this time is how long ago the Big Bang happened, or the age of the Universe. If we substitute $H_o d$ for v from Hubble's Law, we get:

$t = d/v = d/ H_o d = 1/ H_o$

The d's cancel out, so it does not matter which galaxies we choose, and we now see that the age of the Universe is the inverse of Hubble's Constant. Now, let's put in the numbers and find the actual age of the Universe. First, put H_o into standard units—meters and seconds (we will use 71 for the value of Hubble's Constant):

$(71$ km/sec/megaparsec$) \times (1000$ meters/ km$) \times (1$ megaparsec $/3.08 \times 10^{22}$ meters$)$

$= 2.31 \times 10^{-18}$ 1/sec (the units all cancel out except seconds on the bottom)

Now take the inverse of this: $t = 1/ H_o = (1)/(2.31 \times 10^{-18}$ 1/sec$) =$

<u>4.33×10^{17} seconds</u>

Let's express it in years:

$(4.33 \times 10^{17}$ seconds$) \times (1$ year$/3.16 \times 10^7$ seconds$) =$ <u>13.7 billion years</u>

We have found the age of the universe!

Appendix III
Dating Earth's History

Introduction

How do scientists know that the Earth is 4.56 billion years old, or that *Homo erectus* lived in central Asia about 1.75 million years ago, or that the first city-states in Mesopotamia emerged around 5000 years ago? Until very recently, it was nearly impossible to construct accurate chronologies of the past, except through stories and records transmitted from person to person. The ability to determine the age of artifacts like Neanderthal bones, or to find absolute dates for ancient events like the formation of the Earth, has emerged only since the 1950s with the development of radiometric dating techniques.

For hundreds of years before the advent of radiometric dating, there were many attempts to estimate the age of the Earth. Leonardo da Vinci, working in the late 1400s, measured sedimentation rates in the Po River and concluded that nearby rocks must have taken about 200,000 years to form, and therefore the Earth was at least this old. However this early estimate was swept aside by Biblical scholars, who calculated the age of the Earth from the genealogies of Adam and Eve listed in the book of Genesis. The most famous calculation was announced in 1650 by Archbishop James Ussher, who asserted that the Earth—indeed the universe—was made by God on October 22, 4004 BC, and that in the next six days all of life was further created, with humans fully formed. The chronology was widely accepted for the next few hundred years in the Christian west, partly because there was very little evidence to contradict it. The "young Earth" view is still held by some people today, despite overwhelming evidence that the Earth is much older that 6000 years.

Pioneering geologists in the 1700s and 1800s like Nicholas Steno, James Hutton, and William Smith began to study rock layers containing fossils and to understand Earth processes like sedimentation and mountain building. These scientists developed the idea of *relative dating,* based on the layering of deposits. The core idea of relative dating is that the *deeper* the layer is, the *older* it is. This has come to be called the *principle of superposition* and is still a foundational principle of geology and archeology. In the late 1700s, Hutton proposed the *Theory of Uniformitarianism,* which asserts that the geological processes of the past were largely the same as those we see today. It was this idea that made it clear to these early geologists that the Earth must be much older that 6000 years.

In the 1860s physicist Lord Kelvin estimated the age of the Earth to be 24-40 million years by assuming that it was originally molten and then calculating a date of formation based on cooling through conduction and radiation. However, he did not know that the Earth had a powerful internal heat source—radioactive decay—so his estimate gave a very young Earth. By the late 1800s, the phenomenon of radioactive decay was discovered by Henri Becquerel and studied further by Marie and Pierre Curie. In 1903 Ernest Rutherford first proposed that radioactive decay could be used to date rocks, and in 1907 Bertram Boltwood made the first measurement of lead/uranium ratios in very old rocks to conclude that the Earth must be at least 2 billion years old. This was the first "modern" estimate of the age of the Earth and was not that far off the mark.

With the discovery of the neutron by James Chadwick in 1932, the field of nuclear physics emerged, and with it came a more complete understanding of nuclear decay processes. Finally in 1949, Willard Libby from the University of Chicago developed the carbon-14 dating technique and received the Nobel Prize in 1960 for this work. Carbon-14 dating is still one of the most important dating techniques today, but many other radiometric methods have now been refined using other isotopes, including potassium-40, uranium-238, rubidium-87, and uranium-235.

Some Background on Radioactive Decay

What is an isotope? Every element comes in several varieties that are called *isotopes*. Carbon, for example, has three isotopes that occur naturally on Earth. All three of these isotopes have six *protons* in the nucleus—this is what makes it carbon. However *neutrons* also live in the nucleus, and the number of neutrons can vary. Almost all of the carbon atoms found on Earth (about 99%) have six protons and six neutrons in the nucleus and we call this carbon-12. About 1% of carbon on Earth has six protons and *seven* neutrons, so this isotope is called carbon-13; and a tiny amount of naturally occurring carbon is carbon-14, with six protons and *eight* neutrons. The same kind of thing is true for every known element—there are different isotopes of each element because of different numbers of neutrons in the nucleus. This was not understood until the 1930s, when James Chadwick discovered the neutron, which opened the field of Nuclear Physics. For most elements, one or more isotopes will be *unstable*, or *radioactive*. In the case of carbon, it's C-14 that is unstable—it spontaneously turns into nitrogen-14 in a process called radioactive decay.

What is radioactive decay? An isotope that is unstable, or radioactive, does not stay around—it turns into a new isotope of a different element through radioactive decay. In the most common decay processes, a little particle shoots out of the nucleus as the nucleus transforms into a new kind of nucleus, and these particles can be detected with instruments like Geiger Counters, or with photographic film. The shower of particles shooting out of a radioactive sample is commonly called radioactivity, and of course it can be dangerous to living things. The two most common forms of radioactive decay are *alpha decay* and *beta decay*.

In alpha decay the particle that shoots out of the nucleus is called an *alpha particle*. It is relatively large, consisting of two protons and two neutrons (so it is actually a Helium nucleus). When an element undergoes an alpha decay its nucleus loses two protons and two neutrons, thus turning it into a new, lighter element (lighter by two in the periodic table). The original element is often called the "parent" and the new element the "daughter." Here are several examples of an alpha decay:

Uranium-238 ⟶ Thorium-234 + Helium-4
(parent) (daughter) (the alpha particle)

Radium-226 ⟶ Radon-222 + Helium-4
(this is the reaction that produces radon gas inside buildings)

Radon-222 ⟶ Polonium-218 + Helium-4

Notice that the daughter product in the 2nd example, radon-222, becomes the parent in the 3rd example. Radium, radon, and polonium are part of a much larger *decay chain*, as are all radioactive isotopes.

The second common decay process, the beta decay, emits a tiny beta particle, which is simply an electron (this process is more properly called *beta-minus* decay). The beta decay process is the result of a neutron in the nucleus turning into a proton and an electron—thus the nucleus *loses* a neutron, *gains* a proton, and the electron shoots out of the nucleus as the beta particle. Beta decay turns an element into the next heavier element on the periodic table because the nucleus gains a proton. Some examples of beta decay:

Carbon-14 ⟶ Nitrogen-14 + electron (the beta particle) + neutrino

Cesium-137 ⟶ Barium-137 + electron + neutrino

In the early1930s, when pioneering nuclear physicists like Enrico Fermi and Wolfgang Pauli were trying to understand beta decay, something seemed to be missing in the reaction. Pauli proposed that a tiny neutral particle must also be produced, and Fermi named it the *neutrino* even though they could not detect it. Not until 1956 would the first neutrino be detected, and it is now believed that there are six types of neutrinos. They remain very difficult to detect – some can even pass right through the Earth!

Radioactive decay as a clock: half-life. Physicists discovered very early in their exploration of radioactivity that the decay of a particular isotope happened at a very specific rate. For example, the alpha decay of uranium-238 into thorium-234 is *very* slow, while radon- 222 decays into polonium-218 quite quickly. This is most easily expressed as a *half-life*, which is the time it takes for half of a sample to decay.

A radioactive isotope decaying into a new isotope is very much like a block of ice melting into water. Like a radioactive isotope, the melting block of ice has a half-life—it takes a certain amount of time for half of the ice to melt. Of course the half-life of the block of ice depends on the surrounding temperature—is it in a refrigerator, or is it sitting outside on a hot day? However, the half-life of an unstable isotope depends on just one thing: what isotope it is. Uranium-238" melts" very slowly with a half-life of 4.5 billion years, while radon-222, with a half-life of 3.8 days, disappears very quickly. The important point is that the decay of a radioactive isotope can be a very accurate clock.

Equations of Radioactive Decay

Melting ice and radioactive decay are both described by the exponential decay function:

$$\mathbf{M(t) = M_0 e^{-\lambda t}} \quad (1)$$

where **M** is the amount remaining after time **t**, $\mathbf{M_0}$ is the starting amount (at time t = 0), and λ is the *decay constant*. The decay constant and the half-life are very nearly reciprocals, and their relationship can be found be letting $M = (½)\, M_0$ in equation (1) and solving for t. This gives:

$$\textbf{half-life} = \mathbf{(0.693)/\lambda} \qquad \underline{\text{or}} \qquad \mathbf{\lambda = (0.693)/}\textbf{half-life} \quad (2)$$

Each radioactive isotope has a specific decay constant and half-life.

Radioactive decay can also be described in terms of *activity* rather than *amount:*

$$\mathbf{A(t) = A_0 e^{-\lambda t}} \quad (3)$$

The activity is the number of decays per second, which can be measured as the number of particles per second emitted by a sample. The exponential decay of activity simply means that the rate of decay decreases with time, or the *number of particles per second* emitted by a radioactive sample decreases with time.

Example: The activity of a radioactive sample is measured with a Geiger counter and found to be 100,000 decays/second [a decay/second is also named one *Becquerel* (Bq)]. One hour later, the same measurement is made and the activity has decreased to 90,000 Bq. What is the decay constant and half-life of this isotope?

Solution: Use equation (3): $\mathbf{A(t) = A_0 e^{-\lambda t}}$ with A = 90,000 Bq, A_0 = 100,000 Bq, and t = 3600 seconds (1 hour).

Rearranging the equation gives: $A/A_0 = \mathbf{e^{-\lambda t}}$

Exponentiating both sides: $\ln(A/A_0) = -\lambda t$

Putting in numbers: $\ln(90{,}000/100{,}000) = -\lambda\,(3600)$

Simplifying: $-(0.105) = -3600\lambda$ or $\lambda = \underline{2.93 \times 10^{-5}\ \text{sec}^{-1}}$

Then, half-life = $(0.693)/\lambda = (0.693)/2.93 \times 10^{-5}\ \text{sec}^{-1}$ = $\underline{23{,}700 \text{ seconds}}$ = $\underline{6.6 \text{ hrs}}$

Carbon-14 Dating

A small amount of the carbon that makes up the CO_2 in Earth's atmosphere is carbon-14, an unstable isotope with a half-life of 5730 years. So, why should we find *any* C-14 on Earth, if it is constantly disappearing through the process of radioactive decay?

The answer is that C-14 is continually being created high up in Earth's atmosphere as high-energy cosmic ray particles from outer space bombard nitrogen and oxygen molecules. Measurements over the last 50 years show that about *one in a trillion* (more accurately this fraction is 1.3×10^{-12}) carbon atoms in the atmosphere are C-14 and that this fraction is approximately constant over time. In other words, C-14 is replenished by cosmic ray bombardment at the same rate that it disappears by radioactive decay.

Two key assumptions for using C-14 dating are that this balance has remained fairly constant over the last 50,000 years or so, and that it is about the same over all locations on Earth. Neither of these assumptions can be absolutely true, and this is discussed further in the "Problems with C-14 Dating" section following.

When plants take in CO_2 from the atmosphere in the process of photosynthesis, they are absorbing a small amount of C-14. Living plants, therefore, acquire the same concentration of C-14 throughout their cells as the atmosphere, and even though the C-14 is constantly decaying, it is also constantly being replenished as the plant exchanges CO_2 with the atmosphere. When animals eat plants, the same concentration of C-14 is established in the animal, and all the way up the food chain as animals eat other animals. This means that all living things contain the same C-14/C-12 ratio of 1.3×10^{-12}, and this stays constant as long as they are alive.

However when an organism dies, its carbon supply is no longer renewed. C-14 now begins to disappear, and the radioactive clock starts ticking. When an archaeologist digs up something that was once living, like a piece of wood, charcoal, or bone, it can be analyzed to see how much C-14 remains. After one half-life (5,730 years) half of the original C-14 would remain, after two half-lives (11,460 years) one-fourth remains, and so on. After ten half-lives (57,300 years), only about 1/1000 of the original C-14 remains and it becomes almost impossible to detect. For this reason C-14 dating is limited to once-living things that are up to about 50,000 years old.

Example: An archeologist digs up a piece of wood from the buried ruins of an ancient settlement. She sends it off to a radiocarbon dating lab where they manage to extract 107 milligrams of pure carbon from the sample (unfortunately destroying some of the sample in the process). When the carbon is analyzed with an accelerator-based mass spectrometer, it is found to contain 4.17×10^{-11} milligrams of carbon-14. How old is the wood (when did the tree die)?

Solution: We need to compare the amount of C-14 in the sample when the tree died to the amount present today:

Using 1.3×10^{-12} for the ratio of C-14 to C-12 in living things:

(107 milligrams of Carbon) x (1.3×10^{-12}) =

1.39×10^{-10} milligrams of C-14
(this is the amount the *living* tree had)

Now, how long did it take for the amount of C-14 to decay from 1.39×10^{-10} milligrams to 4.17×10^{-11} milligrams?

You could make a quick estimate by knowing that the half-life is 5730 years and seeing the two amounts more simply as 13.9 and 4.17. In one half-life the 13.9 would decrease to 6.95 and in two half-lives it would fall to 3.48. So, roughly, the elapsed time is between one and two half-lives, but closer to two, or between 5730 years and 11,460 years, but closer to 11,460. So you might estimate something around 9,000 years.

To do this accurately, we use the radioactive decay equation (1):

$$\mathbf{M(t) = M_0 e^{-\lambda t}}$$

Solving for t:

$$\mathbf{M/M_0 = e^{-\lambda t}}$$
$$\ln (M/M_0) = -\lambda t$$
$$\mathbf{t = -\ln (M/M_0)/\lambda}$$

From equation (2): $\lambda = (0.693)/\text{half-life (in seconds)} = 3.83 \times 10^{-12}\,s^{-1}$

Then, using $M_0 = 1.39 \times 10^{-10}$ and $M = 4.17 \times 10^{-11}$:

t = **3.14×10^{11} seconds = 10,032 years,** so the tree was last alive about 10,000 years ago.

Note: This date should not be taken to be as precise as it sounds, because of several problems discussed in the next section. The half-life itself (and therefore λ) for C-14 has an uncertainty of +/- 40 years, and this alone creates an uncertainty of about +/- 80 years in the 10,032 years calculated above.

Problems with Carbon-14 Dating

C-14 dating is one of the most powerful dating techniques for relatively young, once-living things, yet it has some inherent problems. To be entirely accurate the following things would all have to be true:

1. The atmosphere has had the same C-14 concentration in the past as now ($C\text{-}14/C\text{-}12 = 1.3 \times 10^{-12}$).
2. All living things acquire and maintain this same overall concentration of C-14.
3. This same concentration is uniform over all locations on Earth.
4. The death of a plant or animal is the point at which it ceases to exchange carbon with the environment.
5. After ceasing to exchange carbon, the C-14 concentration in a plant or animal is only affected by radioactive decay.
6. The sample was handled perfectly, and therefore has no contamination.

In reality, none of these things can be completely true. For example, the C-14 concentration in the atmosphere *has* varied over the last 50,000 years because the effect of cosmic rays has varied; the production rate of C-14 *does* vary at different Earth latitudes; exchange of carbon *can* continue for some time after death; C-14 concentrations after death *can* be affected by chemical, physical, and biological processes; and samples *are* sometimes contaminated and processed incorrectly during analysis.

In recent years, scientists have accumulated data on the history of C-14 concentrations in the atmosphere and how it has varied over time and location. In the last 15 years, calibration tables have been developed to convert *uncalibrated radiocarbon dates* to *calendar dates.* For example, an *uncalibrated* radiocarbon date of 10,000 years ago is now adjusted to become a *calibrated* calendar date of about 12,000 years ago. Reliable calibration tables now go back about 26,000 years and scientists hope to eventually extend these back to 50,000 years.

Also the technology for measuring minute C-14 concentrations is improving all the time. The C-14 dating method has become highly reliable and accurate, but still can only be used to date relatively young samples that were once alive. To date rocks that are millions or billions of years old, radioisotopes with much longer half-lives are needed.

Other Radiometric Dating Techniques

Potassium/Argon—Potassium-40 is found naturally throughout the Earth, and it decays into argon-40 with a half-life of 1.25 $x10^9$ years. In molten rock, the argon daughter product can escape because it is a gas. However once the rock solidifies, the argon-40 cannot escape and becomes trapped in the rock. The ratio of potassium-40 to argon-40 in rocks thus gives the time since the rock solidified. When fossils or artifacts are found in or near a rock layer, their age can also be surmised from the age of the rock. *Useful range:* 200,000 years to 4.5 billion years.

Uranium/Thorium—Uranium-234 decays into thorium-230 with a half-life of 250,000 years. U-234 is water-soluble and found in most sources of water, while thorium is not water-soluble. This method is ideal for dating flowstones like stalactites and stalagmites that form in caves from dripping water. The water contains U-234 and no thorium, but after the flowstone forms with U-234 in it, thorium begins to accumulate. The ratio of thorium to uranium gives the age of the flowstone. This also works for calcium carbonate layers, corals, and travertines because they all solidify from water. Some archeological sites have artifacts that can be dated because they are embedded in or bracketed by carbonate materials. *Useful range:* 1000 years to 500,000 years.

Others—A number of other parent/daughter pairs of isotopes with very long half-lives have been used for geologic dating. As with potassium/argon, and uranium/thorium, the date of formation of a sample is determined by finding the daughter to parent ratio, and knowing the half-life of the parent. Some of the most widely used are:

Uranium-238 to Thorium-234	half-life = 4.5 billion years
Uranium-235 to Protactinium-231	half-life = 704 million years
Rubidium-87 to Strontium-87	half-life = 48.8 billion years
Samarium-147 to Neodymium-143	half-life = 106 billion years

Non-Radiometric Dating Techniques

Dendrochronology (also known as *tree-ring dating*)—The yearly growth cycle of a tree is recorded clearly in growth rings that can be seen when a cross section of the trunk is viewed. Not only can the age of the tree be found by counting rings, but also climate changes are recorded in the rings because drought years produce thinner rings. By studying trees of many different types and ages, scientists can construct a chronology of the area extending as far back as 12,000 years. This is very useful when linked with other dating methods, like C-14. Recently, scientists have begun to construct climate records from the giant New Zealand Kauri tree. It is hoped that these ancient trees will help extend C-14 calibration curves to 50,000 years.

Dating by Radiation Exposure—When certain crystalline materials, like quartz minerals, pottery, teeth, sand grains, or flint are exposed to natural background radiation from the surrounding environment, electrons are freed from the crystal lattice. These electrons become trapped in impurities in the crystal structure and therefore accumulate in the sample. The longer the sample is exposed to the background radiation, the more electrons become trapped. So if the number of trapped electrons can be measured, the sample can be dated. There are presently three methods for measuring the number of trapped electrons:

The electron spin resonance (ESR) method detects the paramagnetic properties, or spin, of the trapped electrons. An advantage of this method is that it does not destroy the sample and can be repeated many times.

The *thermoluminescence* (TL) method uses heat to release trapped electrons. When they are released, they produce a tiny glow of light (luminescence). The intensity of the glow is directly proportional to the number of electrons. This is a one-time procedure.

Optically stimulated luminescence or *photoluminescence* uses laser light to detect the trapped electrons. As with the TL method, the light causes the trapped electrons to glow and this glow can be detected.

All three methods of dating by radiation exposure assume that the present day background radiation level at the site was constant over the dating period, and that the sample was not contaminated by other sources of radiation, like uranium from groundwater. Also, because the radiation from sunlight releases trapped electrons, the sample must be buried and these methods give the time since a sample was buried. *Useful range:* several hundred years to 1,000,000 years.

Obsidian Hydration—Obsidian, or volcanic glass, is the result of the rapid cooling of silica-rich lava and it was often used for tool-making by early humans. When an obsidian surface is freshly exposed, as it would be during the chipping processes of tool-making, the obsidian slowly begins to absorb water from the atmosphere. The amount of water absorbed is proportional to the time since the surface was exposed to the atmosphere. To determine the degree of hydration, a small sample must be cut from the obsidian with a diamond-tipped saw (of course altering the original artifact) and analyzed with a special microscope. Accurate dates can be found only if the temperature and humidity conditions are known, and if the original water content of the obsidian is known. *Useful range:* 100 years to 100,000 years.

Paleomagnetism—The Earth's magnetic field today can be detected using a compass—the needle points north, or toward the Earth's magnetic pole. However scientists have discovered that the north pole has wandered around over the last 10,000 years or so, and that it even *reverses* over longer periods of time. We have now constructed a fairly detailed history of the wanderings of the Earth's magnetic field because it is preserved in the orientations of iron-bearing minerals in rocks. When rocks are hot, the iron-bearing minerals are oriented randomly, but as the rock cools these minerals align themselves with the Earth's field and remain frozen in that position (unless the rock is heated again). The same thing happens with sediments that settle in lakes or the ocean – the iron-bearing minerals align with the Earth's field. Until recently the Earth's magnetic history was known back to about 250 million years ago, but recent advances have extended this to about 4 billion years.

Knowing the history of the Earth's magnetic field, samples can be dated by determining the magnetic orientation in the sample, then locating that in the master history. It is necessary to have a rough idea of the age of the sample (from other dating methods) in order to locate it in the long master history of the Earth's field. This technique is used on core samples extracted from rock formations and also fireplaces. *Useful range:* thousands of years to billions of years.

The Art of Dating: Putting it All Together

It should be clear that dating is both complex and difficult because there are so many factors that cannot be completely controlled or known. It would not be realistic to picture a scientist finding an artifact, sending it off to a lab, then getting back a single firm date for the age of the artifact based on a single dating technique. More typically, scientists study an entire site and the surrounding region, trying to construct a chronology of the area from the layers in and around the site, using as many different dating techniques as possible.

Example 1—Archeology: A Site with Many Layers

Suppose a fictional team of archeologists discovers the site of an ancient city buried in sediment. In the "digging" process, they remove material slowly and carefully (sometimes over many years), working their way downward through a number of layers. The layers show that the site was occupied by people for thousands of years. In an idealized scenario, let's suppose that the following artifacts were discovered by the team in successively deeper layers:

Layer 1— 2 feet down, a Roman coin with the date 160 A.D.
Layer 2— 4 feet down, the remnants of timbers that were roof supports for a building.
Layer 3—7 feet down, a trash dump, containing many animal bones.
Layer 4—8 feet down, a 6-inch thick layer of fine sand with no artifacts.

Layer 5— 9 feet down, part of a large fireplace hearth.
Layer 6— 12 feet down, a 3-inch thick layer of volcanic rock.
Layer 7— 13 feet down, a thick sand and gravel layer, containing pieces of quartz and some obsidian tools.

Each of these layers could be dated in a different way to produce a chronology of the site:

Layer 1— The date on the coin suggests that the site was most recently occupied during Roman times, between 160 and 200 A.D
Layer 2— Tree ring analysis on one of the timbers shows that the tree lived from about 800 B.C. to 200 B.C. Carbon-14 dating on the wood is consistent and shows that the tree last lived between 300 B.C and 100 B.C. so this layer is given a date of 200 B.C.
Layer 3— Bones from the trash dump are C-14 dated to 1000 B.C.
Layer 4— The layer of fine sand suggests that the site was not occupied for a period of time and wind-blown sand covered it. Thermoluminescence analysis of grains of sand produces dates between 1200 and 2000 B.C.
Layer 5— Paleomagnetic dating of the fireplace hearth indicates that it was last used around 2000 B.C.
Layer 6— Apparently a volcanic eruption covered the site with lava, which is perfect for Potassium/Argon dating. This gives a date of 2500 B.C., consistent with the known eruption date of a nearby volcano.
Layer 7— Hydration dating on the obsidian tools gives a date of 5000 B.C. and luminescence dating of the quartz crystals gives a date of 3000 B.C. The two are in conflict, so there remains some uncertainty about when the obsidian tools were made.

Example 2—Geology: Finding the Age of the Earth

Between about 1907 and 1937, a number of scientists estimated the age of the Earth by measuring lead/uranium (Pb/U) ratios in rock samples. This *chemical age* was based on knowing that somehow uranium in the Earth turned into Lead in the radioactive decay process with a half-life of a few billion years. Using chemical processes, they measured the *total* amount of uranium and lead present in a sample, and from this arrived at estimates between half a billion and 4 billion years for the age of the Earth. However not until the 1930s was there an understanding that uranium, lead and other unstable elements came in different types or isotopes that were all part of a much larger decay chain that began with U-238 and ended with Pb-206 (shown next page).

The Uranium-238 Decay Chain

ISOTOPE	HALF-LIFE	DECAY PROCESS
uranium-238	4.5 billion years	
		alpha decay
thorium-234	24.5 days	
		beta decay
protactinium-234	1.14 minutes	
		beta decay
uranium-234	233,000 years	
		alpha decay
thorium-230	83,000 years	
		alpha decay
radium-226	1590 years	
		alpha decay
radon-222	3.83 days	
		alpha decay
polonium-218	3.05 minutes	
		alpha decay
lead-214	26.8 minutes	
		beta decay
bismuth-214	19.7 minutes	
		beta decay
polonium-214	0.00015 seconds	
		alpha decay
lead-210	22 years	
		beta decay
bismuth-210	5 days	
		beta decay
polonium-210	140 days	
		alpha decay
lead-206	STABLE	

Through the 1930s, nuclear physicists developed a much more detailed picture of the decay processes that slowly turned uranium into lead. The isotopes of uranium, lead, and the intermediate members of the decay chain like thorium, polonium, and radium were all identified and half-lives were established so that *isotopic dating* became possible. The technology for separating and identifying minute quantities of isotopes was vastly improved during World War II with work on the atomic bomb. By the end of the war, scientists like Alfred Nier, Arthur Holmes, and Fritz Houtermans began to work on isotopic dating of rocks. They struggled for several years to find the correct amount of "primeval lead" (Pb-204) in rocks, a crucial piece of the isotopic dating puzzle. This very stable isotope is the only isotope of lead *not* produced by radioactive decay. It was present with the formation of the Earth, so its presence needs to be subtracted out from isotopic measurements.

Finally in 1953 Clair Patterson solved the primeval lead problem by analyzing the Canyon Diablo meteorite and measuring the amount of Pb-204 it contained. Meteorites like this formed before the Earth did, along with the early Solar System, and therefore date to just before the formation of the Earth. Using Patterson's data, Houtermans analyzed Earth rocks to arrive at a preliminary age of the Earth of **4.5 +/- 0.3 billion years**. With further study of meteorites and old Earth rocks, Patterson refined this date to **4.55 +/- 0.07 billion years**, and published this value in 1956.

In the fifty years or so since these first accurate determinations of the age of the Earth, there have been many more determinations using much better data. Precision instruments for measuring quantities of isotopes have continually improved, many more meteorites have been analyzed as well as rocks from the Moon, and better values of decay constants have been found. Yet today's best value of **4.55 +/- 0.02 billion years** is not much different than Patterson's 1956 estimate.

Appendix IV
Summaries of Important Hominid Fossils

Sahelanthropus tchadensis Nickname: Toumai

Lived: 6-7 mya

Brain size: about 350 cc

Discovered: 2001 in Chad by Mahamat Adoum, Ahounta Djimdoumalbaye, Fanone Gongdibe, and Alain Beauvilain

Comments: One specimen only. Thought to be the first member of the hominid line, or perhaps the common ancestor of chimps and humans.

Orrorin tugenensis Nickname: Millenium Man

Lived: about 6 mya

Brain Size: unknown, presumed to be about 350 cc

Discovered: 2000 in Kenya by Brigitte Senut and Martin Pickford

Comments: One specimen only. Femur and neck suggest bipedalism. Seems to be more similar to later *Homo* members than to *Australopithecus*.

Ardipithecus kadabba

Lived: 5.8 to 5.2 mya

Brain Size: unknown, presumed to be about 350 cc

Discovered: 2003 in Ethiopia by Yohannes Haile-Selassie, Gen Suwa, and Tim D. White

Comments: One specimen only

Ardipithecus ramadis Nickname: Ardi

Lived: 5.2 to 4.4 mya

Brain Size: unknown, presumed to be about 350 cc

Discovered: 1992 in Ethiopia by Tim D. White's team

Comments: One specimen only. White's team took 16 years to painstakingly reconstruct the crushed specimen, and finally released their findings to the world in 2008. The October 2, 2009 issue of *Science* devoted the cover and eleven articles by 47 different researchers from around the world to Ardi.

Australopithecus anamensis
Lived: 4.4 to 3.9 mya
Brain Size: 350-400 cc
Discovered: 1965 in Kenya by B. Patterson (Harvard). Later specimens in 1994 by Meave Leakey and Alan Walker
Comments: Not designated as *anamensis* until 1994

Australopithecus afarensis Includes Lucy and Selam
Lived: 3.9 to 2.9 mya
Brain Size: 375-500 cc
Discovered: 1973 in Ethiopia by Donald Johanson's team
Comments: Lucy was discovered in 1974 by the Johanson team. She is 3.2 million years old and the most complete skeleton of this era. More recent finds suggest that *afarensis* was much taller and more human-like than previously thought.

Australopithecus africanus Taung Boy is type specimen
Lived: 3.0-2.0 mya
Brain size: 420-500 cc
Discovered: 1924 in Taung, South Africa by Raymond Dart
Comments: Robert Broom found several other *africanus* specimens in South Africa in 1947. It may have been this species that split into *Homo* and *Paranthropus*.

Australopithecus sediba
Lived: about 1.95 mya (based on one site)
Brain size: estimated 420 cc
Discovered: 2008 at Malapa Nature Reserve, South Africa by Matthew Berger, nine-year old son of paleoanthropologist Lee Berger. Berger's team from University of the Witwatersrand has found more specimens and continues work at the site.
Comments: The discoverers consider this a new species, though others suggest it is a variety of *africanus*. Either way it appears to be a transitional hominid between *Australopithecus* and *Homo*, perhaps the immediate predecessor of *Homo habilis*.

Homo habilis Nickname: Handy Man
Lived: 2.4 -1.5 mya
Brain size: 500-700 cc
Discovered: 1960 in Olduvai Gorge, Tanzania by Jonathon and Mary Leakey
Comments: Subsequent finds are scant and date to about 1.9-1.8 million years old. Considered to be the first member of genus *Homo*. Stone tools are found for the first time.

Paranthropus boisei Nickname and type specimen: Zinj

Lived: 2.1 to 1.1 mya

Brain size: 500-530 cc

Discovered: 1959 in Olduvai Gorge, Tanzania by Mary Leakey

Comments: Originally named *Zinjanthropus boisei*, then changed to *Australopithecus boisei*, and now named *Paranthropus boisei* (the 3 species of robust Australopithecines are now placed in the genus *Paranthropus*). *Paranthropus boisei* was the last of the Australopiths.

Homo ergaster

Lived: about 2.0 mya to1.0 mya (?)

Brain size: 700-1000 cc

Discovered: 1975 in Kenya by Bernard Ngeneo

Comments: The most complete specimen is the 1.6 million-year-old Turkana Boy. *Homo ergaster* is widely considered to be the early African version of *Homo erectus*, though some consider both to be *erectus*. *Homo ergaster* probably evolved into *Homo heidelbergensis*.

Homo erectus Includes Java Man and Peking Man

Lived: about 1.9 mya to perhaps 50,000 ya

Brain size: 800-1100 cc

Discovered: 1891 in Java by Eugene Dubois

Comments: This species is generally thought to have originated in Africa as *Homo ergaster*, then it migrated to Central Asia around 1.9 mya, spreading to China, Southeast Asia, and Europe. Non-African specimens are usually called *erectus*, while African specimens are called *ergaster*. Java Man dates to around 700,000 (uncertain because site was disturbed) and Peking Man to about 500,000 years. *Homo erectus* may have reached Spain by 1.2 mya, accounting for Atapuerca specimens.

Homo heidelbergensis Nicknames: Heidelberg Man, Rhodesian Man

Lived: perhaps 1 mya to about 200,000 ya

Brain size: 1100-1350 cc

Discovered: 1907 in Maurer, near Heidelberg, Germany by Otto Schoetensack

Comments: Although *heidelbergensis* was first proposed in 1907, it was largely rejected until recently, but it has now gained wide acceptance as the successor of *ergaster*, and probably the bridge to *Homo sapiens*. Specimen locations include Spain (800,000 ya), Britain (700,000 ya), Ethiopia (600,000 ya), Germany (400,000 ya), Zambia (300,000ya), and Greece (200,000 ya). It is widely believed that the northern (European) version became *Homo neanderthalensis* by about 400,000 ya, while the southern (African) version became *Homo sapiens* by about 200,000 ya.

Homo neanderthalensis Nickname: Neanderthal Man

Lived: about 400,000 ya to 28,000 ya
Brain size: 1350–1750 cc
Discovered: 1829 in Belgium, 1848 in Gibraltar, and 1856 in the Neander Valley near Düsseldorf, Germany (this was the type specimen)
Comments: Long considered our direct ancestor, recent comparisons of Neanderthal DNA to human DNA suggest that the two are parallel species that diverged about 400,000 years ago, yet modern humans of European descent have recently been found to have between 1 and 4% Neanderthal genes. Somewhere along the way there was apparently some interbreeding between humans and Neanderthals. *heidelbergensis* is generally thought to be the probable common ancestor. Many specimens have been found from Gibraltar to Israel, Iraq, and Siberia.

The Denisova Hominid (as yet unnamed)

Lived: perhaps 640,000 ya until 30,000 or 40,000 ya
Brain size: Similar to Neanderthals
Discovered: 2008 in Denisova Cave, Altai Mountains, Russia
Comments: Krause and Pääbo successfully extracted nuclear DNA from a finger bone, and analysis showed that this hominid differed from Neanderthals and humans. They estimated that Denisovans diverged from the Neanderthals about 640,000 years ago and lived until at least 41,000 years ago, which is the age of the finger bone. Genomic comparisons with modern humans show that Melanesians contain 4-6% Denisovan genes, indicating that there was interbreeding sometime in the past.

Homo sapiens idaltu Nickname: Archaic Humans

Lived: Perhaps 250,000 ya to about 70,000 ya
Brain size: 1350–1500 cc
First Discovered: 1967 in Kibish, Ethiopia by Richard Leakey
Comments: It is widely believed that *heidelbergensis* (in Africa) gradually evolved into archaic humans. The specimans Omo I and Omo II date to 195,000 years ago and are the oldest known anatomically modern humans, the ancestor of all later humans. Specimens from Herto, Ethiopia are 160,000 years old. The "Mitochondrial Eve" hypothesis first advanced by Allan Wilson asserts that all living humans can trace their lineage through mitochondrial DNA to a common female ancestor in Africa who lived 200,000 -150,000 years ago. Though anatomically modern, *idaltu* still lacked modern human behavior—symbolic thought and culture would not emerge until 100,000 to 50,000 years ago.

Homo sapiens sapiens Nickname: Modern Humans

Lived: Perhaps 70,000 ya—present

Brain size: 1350-1500 cc

Discovered: 1869 in Southwestern France (Cro-Magnon Man)

Comments: There is widespread debate and uncertainty about when behaviorally modern humans emerged. Clearly by about 40,000 years ago, fully modern humans (Cro-Magnons) lived in Europe, leaving sophisticated tools and stunning cave paintings. However, recent evidence from South Africa suggests that symbolic thought was emerging there by 90,000 years ago or earlier. Stanley Ambrose holds the view that the eruption of Mt. Toba 73,000 years ago created a population bottleneck and it was the African survivors that became fully modern humans who finally spread over the entire Earth starting about 60,000 years ago.

Annotated References

PART ONE

Chapter 1

Bojowald, Martin. 2008. Follow the bouncing universe. *Scientific American* (October).

Cowen, Ron. 2008. State of the universe. *Science News* (15 March). This article summarizes the findings from 5 years of observations of the cosmic microwave background by WMAP.

Chapter 2

Bromm, Volker, et al. 2009. The formation of the first stars and galaxies. *Nature* (7 May).

Cowen, Ron. 2010. Among stars, heavyweight champ. *Science News* (14 August). Astronomers observed the largest known star in 2010 with a mass of 265 solar masses, well outside the theoretical limit of 100 solar masses, so theory will be modified.

Overbye, Dennis. 2003. Scientist at work: Adam Reiss; his prey: dark energy in the cosmic abyss. *New York Times* (February 18). This is a more readable account of the discovery and search for dark energy.

Petit, Charles. 2008. Ultramassive: as big as it gets. *Science News* (25 October).

Reiss, Adam, et al. 1998. Observational evidence from supernovae for an accelerating universe and a cosmological constant. *The Astronomical Journal, 116* (September).

Simcoe, Robert A. 2004. The cosmic web. *American Scientist Online*, 92 (1). <http://www.americanscientist.org/issues/feature/the-cosmic-web/1>.

Chapter 3

Campbell, B., Walker, G. A. H., and Yang, S. 1988. A search for substellar companions to solar-type stars. *Astrophysical Journal* (15 August). This was the pioneering paper describing the radial velocity technique (measuring star wobble) and reporting the first suspected exoplanets. <http://adsbit.harvard.edu/cgi-bin/nph-iarticle_query?bibcode=1988ApJ...331..902C>.

Cowen, Ron. 2011. Liquid acquisition. *Science News* (15 January).

Cowen, Ron. 2011. Planets take shape in embryonic gas clouds. *Science News* (25 March). <http://www.sciencenews.org/view/generic/id/71776/title/Planets_take_shape_in_embryonic_gas_clouds>.

Cowen, Ron. 2011. Spacecraft sees signs of 1200-plus worlds. *Science News* (26 February). <http://www.sciencenews.org/index/generic/activity/view/id/69476/title/ Spacecraft_sees_signs_of_1,200-plus_worlds>.

Huss, Gary R., Tachibana, Shogo, et al. 2008. Injection of short-lived radio nuclides into the early solar system from a faint supernova with mixing fallback. *The Astrophysical Journal, 655*:1382-1387.

Lin, Douglas N. C. 2008. The genesis of planets. *Scientific American* (May).

Matson, John. 2010. Shields up: Magnetized rocks push back origin of earth's magnetic field. *Scientific American Online* (4 May). <http://www.scientificamerican.com/ article.cfm?id=geodynamo-start-up>

Michel, Mayer and Didier, Queloz. 1995. A Jupiter-mass companion to a solar-type star. *Nature* (23 November). The first unquestioned exoplanet orbiting a sun-like star (Cygni 51) was reported by Mayer and Queloz from the Geneva Observatory.

NASA. 2012. <http://lambda.gsfc.nasa.gov/> See for a complete listing of research findings from the WMAP mission and conclusions about the age of the universe.

NASA Press. 1999. Earth's water probably didn't come from comets, Caltech researchers say. <http://neo.jpl.nasa.gov/news/news008.html>.

NASA Science News. 2001. A taste for comet water (18 May). <http://science.nasa.gov/science-news/science-at-nasa/2001/ast18may_1/>.

Portegies Zwart, Simon F. 2009. The long lost siblings of the sun. *Scientific American* (November).

Tachibana, S. and Huss, G. R. 2003. Iron-60 in troilites from an unequilibrated ordinary chondrite and the initial 60Fe/56Fe in the early solar system. *34th Lunar and Planetary Science* (March).

Webb, Stephen. 2002. *If the Universe Is Teeming with Aliens…WHERE IS EVERYBODY?: Fifty Solutions to the Fermi Paradox and the Problem of Extraterrestrial Life.* Las Vegas, NV: Praxis Publishing Ltd.

Wordsworth, Robin D. et al. 2011. Gliese 581d is the first discovered terrestrial-mass exoplanet in the habitable zone. *arXiv.org* at Cornell University Library (5 May). <http://arxiv.org/abs/1105.1031>.

PART TWO

Chapter 4

de Duve, Christian. 1991. *Blueprint for a Cell.* Burlington, NC: Neil Patterson Publishers.

Kauffman, Stuart A. 1993. *Origins of Order: Self Organization and Selection in Evolution.* New York: Oxford University Press.

Kauffman, Stuart A. 2006. Beyond reductionism. *Edge, The Third Culture* (13 November). <http://www.edge.org/3rd_culture/kauffman06/kauffman06_index.html>.

Lee, Rosalind C. et al. 1993. The C-elegans heterochronic gene *lin-4* encodes small RNAs with antisense complementarity to *lin-14. Cel.* (3 December).

Micro-RNA plays important role in mechanisms of human brain development and emergence of some mental diseases. 2006. *Medical News Today.* (25 April). <http://www.medicalnewstoday.com/articles/42141.php>.

Mourelatos, Zissimos. 2008. The seeds of silence. *Nature* (4 September).

Orgel, Lesley E. 1997. The origin of life on earth. <http://eddieting.com/eng/originoflife/orgel.html>.

Tenenbaum, David. 2002. When did life on earth begin? Ask a rock. *Astrobiology Online Magazine* (14 October). <http://www.astrobio.net/exclusive/293/when-did-life-on-earth-begin-ask-a-rock>.

Vines, Gail. 2002. Wonderweed. *New Scientist* (2 December). Vines attributes this estimate to Elliot Meyerowitz of Caltech.

Chapter 5

Collini, Elisabetta, Scholes, Gregory, et al. 2010. Coherently wired light-harvesting in photosynthetic marine algae at ambient temperature. *Nature, 463,* 644-647.

Engel, G.S. et al. 2007. Evidence for wavelike energy transfer through quantum coherence in photosynthetic complexes. *Nature, 446,* 782-786.

PART THREE

Chapter 6

Bryson, Bill. 2003. *A Short History of Nearly Everything.* New York: Broadway Books.

Darwin, Charles. 1871. *The Descent of Man, and Selection in Relation to Sex.* Re-published 2008. New York: Quill Pen Classics.

El Albani, Abderrazak, et al. 2010. Large colonial organisms with coordinated growth in oxygenated environments 2.1 Gyr ago. *Nature* (1 July).

Gould, Stephen J. 1989. *Wonderful Life.* New York: W.W. Norton.

Kump, Lee R. 2008. The rise of atmospheric oxygen. *Nature, 451* (17 January).

Lipps, Jere. 2001. The "star burst" hypothesis for the diversification of the eukaryotes and the geologic record indicate very early origins for all major lineages. Presented at the annual meeting of the Geological Society of America, Boston.

Morris, Simon Conway. 1998. *The Crucible of Creation.* Oxford, UK: Oxford University Press.

Morris, Simon Conway. 2003. *Life's Solution—Inevitable Humans in a Lonely Universe.* Cambridge, UK: Cambridge University Press.

The Pew Research Center for the People and the Press. 2009. Fewer Americans see solid evidence of global warming. (22 October). <http://people-press.org/report/556/global-warming>.

Vines, Gail. 2000. Wonderweed. *New Scientist* (2 December). Vines attributes this estimate to Elliot Meyerowitz of Caltech.

Chapter 7

Algeo, Thomas J. et al. 1995. Late Devonian oceanic anoxic events and biotic crises: "rooted" in the evolution of vascular land plants? *GSA Today,* 5 (3). <ftp://rock.geosociety.org/pub/GSAToday/gt9503.pdf>.

Alvarez, Luis et al. 1980. Extraterrestrial cause for the Cretaceous-Tertiary extinction. *Science* (6 June).

Brooks, T.M., Gittleman, J.L., Pimm, S.L., and Russell, G. J. 1995. The future of biodiversity. *Science, 269,* 347-350.

Falkowski, Paul G. et al. 2005. The rise of oxygen over the past 205 million years and the evolution of large placental mammals. *Science* (30 September). See also <http://www.msnbc.msn.com/id/9536143/-story>.

The Great Dying. 2002. *NASA Science* (28 January). <http://science.nasa.gov/science-news/science-at-nasa/2002/28jan_extinction/>.

Lawton, J.H. and May, R.M. 1995. *Extinction Rates.* London: Oxford University Press.

MacPhee, R.D.E. 1999. *Extinctions in Near Time: Causes, Contexts, and Consequences.* New York: Kluwer Academic Publishers.

Schulte, Peter et al. 2010. The Chicxulub asteroid impact and mass extinction at the cretaceous-paleogene boundary. *Science* (5 March).

Sheehan, Peter M. 2001. The late Ordovician mass extinction. *Annual Review of Earth and Planetary Sciences, 29*, 331-364.
University of California Riverside. 2002. Gondwana split sorts out mammalian evolution. *ScienceDaily* (21 January). <http://www.sciencedaily.com/releases/2002/01/020121090546.htm>.

PART FOUR

Chapter 8

Berger, Lee R. et al. 2010. *Australopithecus sediba*: A new species of homo-like Australopith from South Africa. *Science* (9 April).
Haile-Selassie, Yohannes, Lovejoy, Owen C., et al. 2010. An early *Australopithecus afarensis* postcranium from Woranso-Mille, Ethiopia. *Proceedings of the National Academy of Sciences*.
McPherron, Shannon P. 2010. Evidence for stone-tool-assisted consumption of animal tissues before 3.39 million years ago at Dikika, Ethiopa. *Nature, 466* (7308), 857-860.
Raichlen, David A. et al. 2010. Laetoli footprints preserve earliest direct evidence of human-like bipedal biomechanics. *PLoS One* (22 March). <http://www.ncbi.nlm.nih.gov/pmc/articles/PMC2842428/>.
Science Magazine. 2009. October 2 issue devoted the cover and eleven articles to *Ardipithecus ramidus*, authored by 47 different researchers from around the world including Tim White, C. Owen Lovejoy, and Yohannes Haile-Selassie.
Stone, Richard. 2009. Paleoanthropology: Still seeking Peking man. *Science* (3 July).
Thorpe, S. K. S., Holder, R. L., and Crompton, R. H. 2007. Origin of human bipedalism as an adaptation for locomotion on flexible branches. *Science* (1 June).
Wong, Kate. 2010. The first butchers. *Scientific American* (October).

Chapter 9

Bower, Bruce. 2011. Oldest hand axes discovered. *Science News* (8 October).
Bryson, Bill. 2003. *A Short History of Nearly Everything*. New York: Broadway Books.
Carbonell, Eudald et al. 2008. The first European? *Nature* (27 March).
Feldman, Marcus, Cavalli, Luigi L., Myers, Richard M., et al. 2008. Worldwide human relationships inferred from genome-wide patterns of variation. *Science* (22 February).
Hanihara, Tsunehiko, Manica, Andrea, et al. 2007. The effect of ancient population bottlenecks on human phenotypic variation. *Nature* (19 July).
Krings, Matthias, Pääbo, Svante, et al. 1999. DNA sequence of the mitochondrial hypervariable region II from the Neanderthal type specimen. *PNAS* (11 May).
Lisiecki, Lorraine E. and Raymo, Maureen E. 2005. A Pliocene-Pleistocene stack of 57 globally distributed benthic $\delta\,^{18}O$ records. *Paleoceanography, 20*.
Lordkipanidze, David et al. 2007. Postcranial evidence from early *Homo* from Dmanisi, Georgia. *Nature* (20 December). <http://www.dmanisi.org.ge/majordiscoveries.html>.
National Research Council (U.S.) Committee. 2006. Surface temperature reconstructions for the last 2,000 years. *National Academies Press*.
Pääbo, Svante et al. 2006. Analysis of one million base pairs of Neanderthal DNA. *Nature* (16 November).
Pääbo, Svante, Pritchard, Jonathan K., Rubin, Edward M., et al. 2006. Sequencing and analysis of Neanderthal genomic DNA. *Science* (17 November).

Pääbo, Svante et al. 2007. The derived *FOXP2* variant of modern humans was shared with Neanderthals. *Current Biology* (6 November).
Pääbo, Svante, Reich, David, et al. 2010. A draft sequence of the Neanderthal genome. *Science* (7 May).
Potts, Richard. 1996. Evolution and climate variability. *Science* (16 August).
Reich, David, Viola, Bence, Pääbo, Svante, et al. 2010. Genetic history of an archaic hominin group from Denisova Cave in Siberia. *Nature* (23 December).
Rose, Mark. 1997. A new species? *Archaeology Online News* (29 July).
Saey, Tina Hesman. 2010. Modern people carry around Neanderthal DNA, genome reveals. *Science News* (5 June).
Wong, Kate. 2009. Twilight of the Neandertals. *Scientific American* (August).

PART FIVE

Chapter 10

Ambrose, Stanley. 1998. Late Pleistocene human population bottlenecks, volcanic winter, and differentiation of modern humans. *Journal of Human Evolution* (June).
Balter, Michael. 2009. Early start for human art? Ochre may revise timeline. *Science* (30 January).
Behar, Doron M. et al. 2008. The dawn of human matrilineal diversity. *American Journal of Human Genetics* (24 April).
Bower, Bruce. 2002. Chinese roots: Skull may complicate human-origins debate. *Science News* (21 December).
Cann, Rebecca L., Stoneking, Mark, and Wilson, Allan C. 1987. Mitochondrial DNA and human evolution. *Nature* (1 January).
Cavalli-Sforza, L.L., Underhill, P.A., et al. 1996. Geographic clustering of human Y-chromosome haplotypes. *Annals of Human Genetics* (September).
Curtis W. Marean et al. 2007. Early human use of marine resources and pigment in South Africa during the Middle Pleistocene. *Nature* (18 October).
Diamond, Jared. 1992. *The Third Chimpanzee*. New York: Harper Collins.
Donnelly, Peter et al. 2000. The mutation rate in the human mtDNA control region. *The American Journal of Human Genetics* (1 May).
Green, Richard E., Reich, David, Pääbo, Svante, et al. 2010. A draft sequence of the Neandertal genome. *Science* (7 May).
Ingman, Max, Kaessmann, Henrik, Pääbo, Svante, and Gyllensten, Ulf. 2000. Mitochondrial genome variation and the origin of modern humans. *Nature* (7 December).
Klein, Richard G. and Edgar, Blake. 2002. *The Dawn of Human Culture*. New York: Wiley and Sons.
Petraglia, Michael et al. 2007. Middle Paleolithic assemblages from the Indian subcontinent before and after the Toba super-eruption. *Science* (5 July).
Richerson, Peter J. and Boyd, Robert. 2006. *Not By Genes Alone*, 5. Chicago: University of Chicago Press.
Scholz, Christopher A. et al. 2007. East African megadroughts between 135 and 75 thousand years ago and bearing on early-modern human origins. *PNAS* (16 October).
Stix, Gary. 2008. Traces of a distant past. *Scientific American* (July).

Stoneking, Mark and Pakendorf, Brigitte. 2005. Mitochondrial DNA and human evolution. *Annual Review of Genomics and Human Genetics*, 6, 165-183.
Vanhaeren, Marian, d'Errico, Francesco, et al. 2006. Middle Paleolithic shell beads in Israel and Algeria. *Science* (23 June).
Weber, George. 2004. Toba Volcano. <http://www.andaman.org/org/BOOK/originals/Weber-Toba/textr.htm>. An excellent discussion of the Toba event and human evolution.

Chapter 11

Balter, Michael. 2009. Whence the first Americans? *ScienceNow Daily News* (8 January).
Bar-Yosef, Ofer. 1998. The Natufian culture in the Levant, threshold to the origins of agriculture. *Evolutionary Anthropology*, 6, 159-177.
Cavalli, Sforza L.L., Underhill, P.A., et al. 1996. Geographic clustering of human Y-chromosome haplotypes. *Annals of Human Genetics* (September).
Coatsworth, John. 1996. Presidential Address: Welfare. *The American Historical Review* (February).
Cruciani, Fulvio et al. 2011. A revised root for the human Y chromosomal phylogenetic tree: the origin of patrilineal diversity in Africa. *The American Journal of Human Genetics* (10 June).
Diamond, Jared. 1999. *Guns, Germs, and Steel*, 36. New York: Norton.
Diamond, Jared. 2005. *Collapse: How Societies Choose to Fail or Succeed.* New York: Viking.
Dillehay, Tom D. et al. 2007. Preceramic adoption of peanut, squash, and cotton in northern Peru. *Science* (29 June).
Dillehay, Tom D. et al. 2008. Monte Verde: Seaweed, food, medicine, and the peopling of South America. *Science* (9 May).
Finlayson, Bill and Kuijt, Ian. 2009. Evidence for food storage and predomestication granaries 11,000 years ago in the Jordan Valley. *PNAS* (July 7).
First Americans arrived as two separate migrations, according to new genetic evidence. 2009. *ScienceDaily* (21 January). <http://www.sciencedaily.com/releases/2009/01/090108121618.htm>.
Gilbert, M, Thomas, B., et al. 2008. DNA from pre-Clovis human coprolites in Oregon, North America. *Science* (9 May).
Krause, Johannes et al. 2007. The derived FOXP2 variant of modern humans was shared by the Neanderthals. *Current Biology* (6 November).
Pääbo, Svante, Reich, David, et al. 2010. A draft sequence of the Neanderthal genome. *Science* (7 May).
Pakendorf, Brigitte and Stoneking, Mark. 2005. Mitochondrial DNA and human evolution. *Annual Review of Genomics and Human Genetics* 6, 165-183.
Perego, Ugo A. et al. 2009. Distinctive Paleo-Indian migration routes from Beringia marked by two rare mtDNA haplogroups. *Current Biology* (13 January).
Pringle, Heather. 2011. The 1st Americans. *Scientific American* (November).

Chapter 12

Christian, David. 2004. *Maps of Time: An Introduction to Big History*. Berkeley CA: University of California Press.

Harris, Marvin. 1974. *Cannibals and Kings: The Origins of Cultures*. New York: Random House.
Lawler, Andrew. 2007. Archaeology: ancient writing or modern fakery? *Science* (3 August).
Leakey, Richard E. and Lewin, Roger. 1991. *Origins*. New York: Penguin Books.
Ur, Jason A. et al. 1974. Early urban development in the Near East. *Science* (31 August).

Index

C

D

E

F

G

H

I

J

K

L

M

N

Q

R

S

T

U

V

W

Y

Z